MW00514231

Methods for Solving Process Plant Problems

Methods for Solving Process Plant Problems

by Joel O. Hougen, Ph.D.

Resources for Measurement and Control Series

Printed in the United States of America.

ISA
67 Alexander Drive
P.O. Box 12277
Research Triangle Park
North Carolina 27709

Library of Congress Cataloging in Publication Data

Hougen, Joel O.
Methods for solving process plant problems / by Joel Hougen.
p. cm.
Includes bibliographical references and index.
ISBN 1-55617-539-6
1. Chemical process control. I. Title.
TP155.75.H676 1996
660'.281'0724--dc20 96-3197
CIP

Acknowledgments

Two persons deserve mention for providing the opportunity and motivation for the preparation of this book. One is my brother, the late Professor Olaf A. Hougen, the other is my wife, Alma Thorsheim Hougen.

I was one of many fortunate recipients of Olaf's generous and steadfast assistance, encouragement and direction, which for me occurred prior to and throughout student days and much of my professional career.

If my obligations to my wife are acknowledged last it is only because they are the greatest of all, since in large tasks and small, from the beginning to the end, she has helped. She kept the family fabric from unraveling during my numerous and prolonged absences required in conducting these studies, and her patience and forbearance permitted expending precious retirement time in the preparation of this book.

One cannot receive my thanks directly, the other receives it daily in spoken and unspoken expressions of gratitude.

Contents

Preface

Opportunities to participate in the transfer of technology from one discipline to another are rare. It was my good fortune to be part of such an event. This book is a partial documentation of my efforts to apply some concepts of "system engineering" to the chemical processing industry.

The casual scanning of Brown and Campbell's book, *Principles of Servomechanisms* (1948) may have stimulated my desire to explore possible applications to chemical processing systems. Subsequent visits with Professor Campbell, which on one occasion included Norbert Wiener, were both inspirational and informative. Later Professor Campbell's book entitled *Process Dynamics*, published posthumously in 1958 with the help of the late Page Buckley of DuPont, gave me encouragement and direction.

A chance encounter in 1952 with Dr. Sidney Lees, at that time at MIT, provided the principal motivation for my desire to explore the possibilities of obtaining and applying frequency response information for process diagnosis and design of process control systems. It was he who introduced me to both the pulse testing technique for exciting the dynamics of systems and the computational procedures for extracting frequency response information from time histories of such tests.

The Frequency Response Symposium, December 1-2, 1953, chaired by Prof. Rufus Oldenberger, made me aware of what was transpiring across a wide spectrum of engineering activity, including chemical engineering. The paper by Smith and Triplett in that symposium, illustrating the use of pulse methods for exciting the dynamic response of aircraft, was especially significant.

There followed years of study, primitive and groping exploratory experimentation, and continued consultations with academic and industrial personnel at conferences and during visits to industrial and

University sites. Together with Robert Walsh, a student in Chemical Engineering at Rensselaer Polytechnic Institute during the years 1950-52, I perused the monumental work of Draper, McKay and Lees,[1] extending our knowledge of applied mathematics, system analysis, and experimental methods.

In the 1940s the dynamics of chemical processes were beginning to become of concern to practicing chemical engineers. Shell Development in the U.S. and Holland, and British Petroleum in England were the first to initiate vigorous programs aimed at using dynamic response information obtained from full-scale processing units. Experimental results from such facilities began to appear in the literature. Some of this early history is mentioned in the publication listed below.[2]

Initial experimental work with processes used direct sinusoidal forcing as a method of obtaining frequency response data. The task of creating the mechanisms for introducing sinusoidal forcing, usually by appropriate motion of ponderous valves, proved to be a major undertaking. In addition, the time required to obtain data covering a satisfactory range of frequency was soon recognized as a major impediment. There had to be a better way!

The pulse technique clearly appeared to be attractive for obtaining frequency response information about chemical processes, and in 1956 its applicability was demonstrated; the results obtained from pulse tests of a small industrial heat exchanger compared favorably to those obtained from direct sinusoidal forcing.[3] The testing of industrial processes, and subsequent reduction of test data to useful form, now appeared feasible.

Fortunately, prior to engaging in the testing of large-scale processes, I had the opportunity to experiment with small-scale processes in a laboratory environment. A wide selection of sensing components was also tested and measurement skills developed. Data reduction routines and the pulse testing technique itself were verified.[4] Thus, when experimenting with full-scale systems, attention could be fully focused on the practical problems associated with acquiring acceptable data in the unforgiving environment of manufacturing plant facilities. The assistance of Technician Oscar R. Martin was invaluable at this crucial point, as were the enthusiastic support and constructive inputs from Prof. G. E. Dreifke of St. Louis University and Dr. Sidney Lees.

---

1. Draper, C. S., W. McKay, and S. Lees. 1952, 1953, 1955. Instrument Engineering, McGraw-Hill Book Company, Inc. Three volumes.
2. J. O. Hougen. 1964. Experiences and Experiments with Process Dynamics, Chemical Engineering Progress Monograph Series, Vol. 60, No. 4.
3. Lees, S. and J. O. Hougen. 1956. Ind. Eng. Chem., 48, p. 1064.
4. Dreifke, G. E. 1961. "Effects of Input Pulse Shape and Width on Accuracy of Dynamic System Analysis from Experimental Pulse Data." A doctorate thesis presented to Washington University, Sever Institute of Technology, St. Louis, Mo.

It is with this background, drawing on the contributions, knowledge, experiences, and encouragement of others, that the work described here has its foundation.

Introduction

The experimental studies presented were all conducted a number of years ago. Some facilities have long since completely vanished, others, if existing, have been drastically modified and hence bear little relation to the original facilities. Although we do not believe that any information of a proprietary nature is disclosed, permission has been obtained where any doubt existed. The studies included here are typical examples representing about a third of the studies carried out by the author.

One may look in vain for explicit evidence of economic benefits derived from these studies. Although obvious or implied, this is lacking because sufficient cost data were not always available or have been considered proprietary. We must leave it to the reader to judge the merits of these studies based on the evidence presented.

Because industrial sensors were inadequate, in all these studies the installation of special sensors or transducers was required for measuring important variables. In all cases special data recorders were used, these being multichannel oscillographs. With these data acquisition systems signal changes as small as 2-3 microvolts could be detected and recorded at any reasonable rate. Since all transducers and sensors were calibrated, accuracy sufficient for the intended purposes was always assured.

While it might seem that modern industrial data systems could provide data of comparable quality, evidence of such does not appear in current literature. A possible handicap may lie in the inability to dedicate the data system to a single processing entity, thus enabling the acquisition of each channel of data in a virtually continuous manner and with the necessary sensitivity. During the progress of these studies, field devices such as sensors, transmitters, and valves, have markedly improved. Nonetheless, some items may yet not be completely satisfactory as complements to digital control systems nor for precise experimental studies.[5]

The circuits associated with the sensors or transducers used in these studies were, in all cases, simple. Typical circuits are shown in an earlier publication.[6] The final assembled data systems were not excessively expensive, the major items being the recording oscillographs. These were of high quality and were provided with a wide range of adjustments for amplification, signal suppression, and chart speed.

---

5. Solley, L. W. 1995. "Spend Your Money in the Basement." 50th Annual Symposium on Instrumentation for the Process Industries, Texas A&M University.
6. Hougen, J. O. Measurements and Control Applications, Chapter 7. ISA. Out of print.

Signal corruption, or noise, was never a serious problem even though some low-level dc signals were transmitted through 500 feet of signal cable. Grounding and shielding was important and required care in assembling and locating the data system. Signal noise could always be eliminated, or sufficiently suppressed, by the proper selection of a capacitor placed across the input of the dc amplifiers.

During the testing phase safety was always a consideration, in respect to both the data system and execution of the testing protocol. Plant personnel were responsible for installation of sensors and introducing changes in process variables. Cooperation of operators and engineering personnel was always solicited.

One objective of this book is to bring to the attention of teachers and students the relative simplicity of solving industrial process problems through experimentation. Using examples taken from studies of real processes provides unequivocal evidence of both validity and practicality. If chemical engineering students would have at least a modest introduction to the art and science of measurement and some exposure to method of data acquisition, data reduction and interpretation, and experimental techniques, they would be better prepared to deal with problems to which they certainly will be exposed. They could then approach a problem with confidence and with the assurance that if appropriate data are procured a solution can be found. Currently, too many young engineers begin industrial employment having had either negligible or negative experiences in university laboratory courses. As a result, they are not capable of performing simple engineering measurements with acceptable precision, nor do they ever experience the joy of discovery. They are not prepared to judge the quality of the data presented to them by data systems, the details with which they may not be familiar. Nor are they likely to accept responsibility for the performance of these data systems. All too frequently knowledge of the details of these systems are assumed to be the responsibility of plant instrument personnel or others whose direct responsibility does not necessarily include solving processing problems. I would be more than rewarded for my efforts if this book proves useful to the curious teacher, student, or engineer who desires to learn more about the processes with which they should be concerned.

The design of processing plants and their control systems may be accomplished utilizing only a modicum of data, provided the important relationships could be derived completely from first principles. For example, if the mechanisms of mass, heat, and momentum transfer, and kinetic relationships were known in sufficient detail, the design of an exothermic reactor and ancillary apparatus could be achieved. Even with a relatively small amount of fundamental information, this might be accomplished and performance predicted.

However, whereas it may be possible to derive formal expressions for the relationships describing physical and chemical processes and reactions, it is so extremely difficult to make satisfactory predictions that precise determinations of the performance of actual processing systems must usually be accomplished by experimentation. From experimental data empirical models can be developed that are usually simpler and frequently more tractable than those derived from theoretical analyses alone. The long history of successful engineering enterprise bears testimony to the need for and usefulness of empirical approaches to achieve solutions to difficult problems. Any of the processes studied here could serve as an example of those that were designed and operated profitably before all technical and scientific details were available.

While improvement in control was a major motivation for the studies described, more than dynamic response information was obtained in all cases. Not only were transient tests performed but also a considerable amount of steady-state data obtained. From the latter, material and energy balances were computed. In one case, these balances closed to within about 3%, demonstrating that this degree of accounting for both material and energy is not only possible, but is a requirement, for solving some problems.

All studies conducted by this author demonstrated that the processes were conservative. Energy does not oscillate within a system, as can be observed in many electrical and mechanical systems. Heat and momentum transfer mechanisms are unidirectional. Positive feedback does not occur within a given processing entity. An exothermic reactor, the effluent of which exchanges thermal energy with the feed in a heat exchanger, can, of course, exhibit positive feedback characteristics, but the reactor and heat exchanger by themselves do not.

Test results also testify that the processes studied are adequately described by deterministic, not stochastic, designations. At given values of independent variables the dependent variables reproduce themselves, sometimes including chemical compositions.

Many processes act as integrators, in fact any process in which the inventory of material or energy changes as a result of a change in either an input or output variable can exhibit this property. This behavior is commonly overlooked when selected control algorithms for process control.

In summary, the studies presented here emphasize the need for a knowledge of details: details of processes, details of individual processing items, details of sensors, and details of control components. In all cases where successful problem solving has occurred this has been true.

About the Author

For over 40 years, Joel Hougen has devoted his professional life to chemical engineering in industry and higher education. He is well known for his ability to identify and develop solutions to problems encountered in the chemical processing industry. This success is attributable to the precision, accuracy and sensitivity of the data systems employed.

His undergraduate studies in Chemical Engineering were completed at the University of Wisconsin in 1936 and later he earned Masters and Ph.D. degrees from the University of Minnesota. His professional career has been equally divided among industry, teaching, and consulting.

Dr. Hougen has taught Chemical Engineering at the University of Minnesota, UCLA, University of Illinois, Rensselaer Polytechnic Institute, St. Louis University, and The University of Texas at Austin. At the latter he held the Alcoa Professorship. He has presented short courses sponsored by the American Institute of Chemical Engineers and The University of Texas. His contributions to the professional literature include over 50 publications and two books.

Since 1978 he has been a Fellow in the American Institute of Chemical Engineers and for 32 years he has been a member of ISA.

Upon retirement he was appointed Professor Emeritus and resides in Austin, Texas.

1 Electric Furnace Pressure Control

A Neophyte Ventures Forth

Imagine being caught in a snow storm in a mountainous region accompanied by a group of experienced, but fair-weather, hikers. Although familiar with the terrain on other occasions, when covered with snow the landscape becomes a trackless wasteland with all landmarks obscured. Yet some of the party may be quick to issue instructions along with hasty suggestions for action and explicit directions to a safe haven.

This situation is analogous to that in which the author found himself in his first exposure to an industrial problem that was to serve as a test of his ability to apply the principles of "system engineering" to the chemical industry. He had, however, been employed in industry long enough to know that information available to fellow engineers was often lacking in precision and accuracy so that, despite their best efforts, correct diagnoses and remedial efforts were often disappointing.

Nevertheless, possessed by an insatiable desire to determine if the theories of servomechanism design could be applied to process control systems, the author ventured forth into the forbidding wilderness. There were few guidelines, no trail had been blazed, and numerous pitfalls were to be encountered.

Fortunately he was not alone. Good fortune provided a remarkably able technician, skilled in electronics, familiar with transducers and their application, and conversant with analog data acquisition systems. Plant personnel were always cooperative and friendly and, in all instances, contributed to the success of the studies.

The material presented herein has been abstracted from reports documenting, in great detail, all facets of each study. These included large amounts of experimental data procured from careful measurements of temperatures, pressures, flow rates, and other variables from many parts of the process being studied. Only a portion is presented herein, principally that bearing on the diagnoses of, and solutions to, the particular processing and control problems encountered.

The Process

This first experimental study describes the work undertaken to solve a problem of long standing, the essence of which was to devise a system to control, very precisely, the pressure in the vapor space of an electric furnace used to produce phosphorus.

In such furnaces elemental phosphorus is produced by the reaction of a mixture of phosphorus ore, coke, and silica-bearing material in the vicinity of an electric arc, where the temperature is in the neighborhood of 2500°F. The "burden", or mixture of solids, is enclosed in a graphite-lined cylindrical vessel about 25 ft. in diameter and 20 ft. high with a ceramic top. Three electrodes about 40 in. in diameter enter the top, through seals, and terminate within the burden. The seals surrounding the electrodes prevent gases from escaping except when excessive pressure develops. Electrical energy enters the electrodes creating an arc that heats the burden. At the extreme temperature in the vicinity of the arc the highly endothermic reaction occurs, producing gaseous P_4, CO, and molten slag. The latter is periodically withdrawn from the bottom of the furnace.

The feed, about the size of peas, consists of a mixture of coke, siliceous material, and granules of ore, the latter prepared by crushing the clinker obtained by heating the phosphorus-bearing ore to the point of incipient fusion. The feed, from an overhead hopper, enters the furnace through chutes that direct the material to the vicinity of each electrode.

Power consumption must usually be maintained within rather close limits because severe penalties for overloads may be incurred. Slow adjustments in power consumption may be accomplished by raising or lowering the electrodes by using motor-driven winches.

The processing scheme, depicted in Figure 1-1, occupied an extensive area and consisted of massive processing items with several hundred feet of large-diameter interconnected piping. At one end the gaseous phosphorus was produced, at the other the phosphorus-free gas, largely CO, was delivered to a manifold into which gases from other processes may discharge. At an intermediate point the phosphorus vapor is condensed, removed, and eventually stored under water. Gas from the low-pressure manifold was forced into another manifold maintained at a pressure sufficient to supply other processes with this fuel.

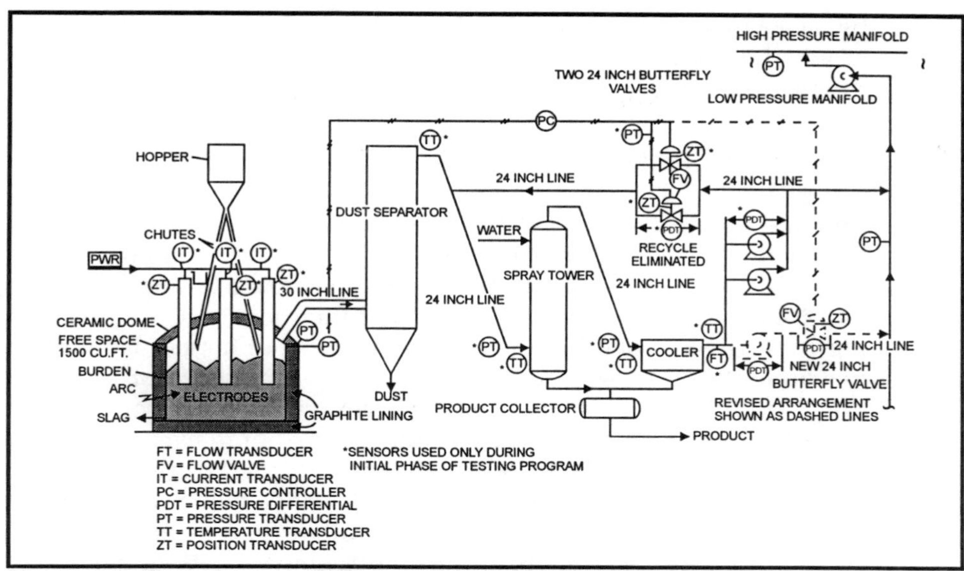

Figure 1-1. Flow diagram of process for producing phosphorus

The Problem

The major problem that existed was the inability to control the pressure in the vapor space within the furnace. Because the ceramic cover could not be made impervious, P_4 gas could escape when the pressure exceeded atmospheric. Upon exposure to air the gas would ignite creating a dense cloud of P_2O_5. On the other hand, if pressure within the furnace were below atmospheric, air would enter the furnace, producing P_2O_5 within the process. When in contact with water in the scrubbers, H_2PO_4 would form, causing corrosion of equipment in addition to loss of product. A pressure slightly above atmospheric, about 1/4 in. of water, was desired.

The effluent gas was conducted through a ceramic-lined, 30-inch transfer line and an electrostatic precipitator about 12 feet on a side and 120 feet high. In the latter, dust was removed, descending to the bottom to be periodically withdrawn. Phosphorus was condensed in a vertical vessel into which water was sprayed. This was followed by a cooler to remove residual phosphorus.

Free space above the burden in the furnace was about 1500 cu. ft.; the volume of the dust precipitator, scrubber, interconnecting pipe, and other vessels was about 7,000 cu. ft. These estimated volumes are mentioned because of their role in determining the transient response of pressure caused by disturbances.

Two large centrifugal fans, operating in parallel, were used to handle the gas from the furnace. With the intent of controlling the furnace pressure, part of the effluent from these fans was directed back to the process through a pair of 24-inch butterfly valves. Furnace pressure was measured by a sensor-recorder-controller located near the furnace outlet, the output of which supplied a variable pressure to the diaphragms of the valves in the recycle lines.

Measurements and the Data Acquisition System

The first task was to determine the sources or causes and the magnitude of the rapid changes in furnace pressure that occurred. Eighteen signals were measured with sensor outputs directed to a multichannel recording oscillograph. (These measurements and their location are indicated in Figure 1-1). Several other measurements were obtained using plant instrumentation or special test instruments. Since pressure disturbances might occur because of changes in power input or electrode motion, measurements of total power consumed, current to each electrode, and displacement of each electrode were also recorded during preliminary tests. From the motion detectors, both electrode velocity and acceleration could be obtained. Power consumed in the furnace was measured with a current transformer and electrode displacements with linear potentiometers attached to each electrode.

A strain-gage-type pressure transducer, capable of sensing pressure changes as small as 1/100 in. WC, was installed to measure furnace pressure at the furnace outlet. Other pressures and pressure differentials were also measured with similar transducers. Since highly transient pressure signals were suspected, transducers were selected which could faithfully follow signals changing as rapidly as 100 Hz. To protect the interior of these transducers from the phosphorus-bearing atmosphere, nitrogen at a low and constant pressure was admitted through small restrictions to all transducer leads when measurements were not being procured. Sensor outputs were conducted to the recording oscillograph through well-grounded, shielded signal cables. The oscillograph pens could faithfully follow 100 Hz signals. Circuits were also arranged by which input signals could be suppressed, or nulled out, so that perturbations around any desired level could be highly amplified. Suppression of the input signals was accomplished by calibrated multi-turn potentiometers in the signal conditioning circuits.

Several Pitot-Venturi-type flow sensors were also installed to measure flow rates of gas. However, because solid and liquid material entrained in the gas streams could impair the operation of these sensors, their outputs were considered unreliable.

Preliminary Observations

In a matter of hours the major cause of pressure changes in the furnace was discovered, this being variations in the electric power consumed. An oscillographic record of power obtained early in this study is presented in the inset shown in Figure 1-2, from which both the amplitude and frequency of pressure changes can be estimated. The power input was rarely constant for more than a few seconds, and the amplitude of the induced pressure changes was usually several inches of water, with some variations as large 8 inches. On occasion furnace pressure changes occurred because of irregularities in the rate of feed descending into the furnace, but these events were less frequent and quite impossible to predict or eliminate, although they could cause pressure changes as great as 6 inches of water or more.

Upon inspecting the installed sensor of furnace pressure, one reason for poor pressure control immediately became clear. This device received the pressure signal through a tube extending beneath the open end of a bell floating in mercury. Changes in pressure caused the bell to move vertically, this motion being transmitted to the recording and pneumatic transmission mechanisms. This bell was a massive element, weighing, perhaps, 400 grams. Figure 1-2 shows the pneumatic output response of this device to a pulse-like input. This instrument was obviously entirely unsuited for measuring the rapidly fluctuating furnace pressure.

Other components in the control system were examined next. The massive (24 in.) butterfly valves located in the recycle lines were immediately suspect. The quality of response of these motor valves, which were not provided with positioners, could be judged by merely observing their responses resulting from abrupt changes in the pneumatic pressure applied to the diaphragms of their actuators. This procedure clearly demonstrated that the valves were unsatisfactory, since movement would not occur for at least a second after a change in diaphragm pressure, and subsequent stem rotation was very slow. The response of valve stem displacement to a pulse change in input pressure to the pressure sensor is also shown in Figure 1-2.

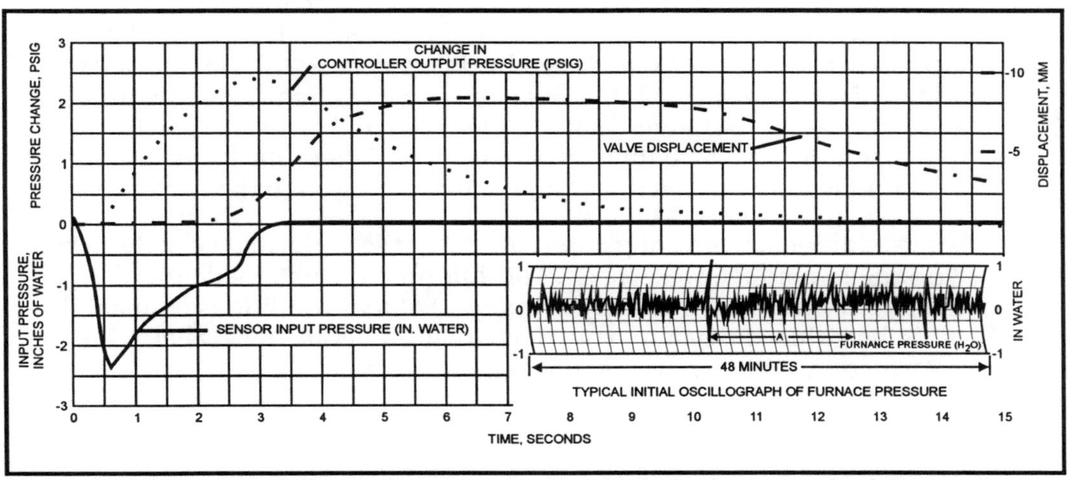

Figure 1-2. Response of plant pressure sensor and control valve to a pulse input, and oscillograph record of typical pressure variations of furnace pressure

Because recycling gas back to a point upstream of the water scrubber as a means of controlling furnace pressure appeared to be a questionable practice, a series of observations was obtained under open-loop conditions, and with the recycle valves closed. Except for a small amount of gas admitted to the system at various points as purges, the exhauster fans were now handling only the gas produced in the furnace. Data obtained under these conditions, as power was changed from zero to 29 megawatts, are shown in Table 1-1. These data demonstrated that the fans were operating well below their capacity as evidenced by the relatively high pressure developed (shown below and in Figure 1-3).

Table 1-1. Open-Loop Performance of Fans at Various Power Levels with Recycle Valves Closed

Power, megawatts	Furnace Pressure, in. water	ΔP Across Fans, in. water	ΔP Across Valves, in. water	Controller Output, psi	Valve Displacement, Degrees
0	.15	14.0	10.75	4.12	22.0
18	.38	14.2	8.00	6.54	34.0
20	.40	14.2	7.50	6.84	36.0
23	.65	13.9	6.50	7.29	38.5
25	.85	13.7	5.00	7.66	41.5
29	.25	13.4	4.13	8.24	45.0

This information demonstrated that the fans, not having to handle the recycled gas, were certainly not overloaded; a single fan, if not required to handle recycled gas, would probably be adequate.

Further information, obtained when the fans were handling recycle gas, suggested that they were overloaded. As is well known, as the flow rate through a centrifugal blower increases, the pressure developed also increases, reaches a peak, and then begins to decrease. When the latter occurs, the fan becomes decreasingly effective in attenuating disturbances appearing at its discharge. This was observed in this installation by noting that disturbances originating from either the high- or the low-pressure gas manifold readily migrated back to the furnace through the recycle path.

Some additional interesting observations on the performance of these fans, working in parallel, were made. This information, obtained with the existing control system in service, suggested that conditions of instability could be induced when two overloaded fans operate in parallel. (Data are presented in Table 1-2 and Figure 1-4.)

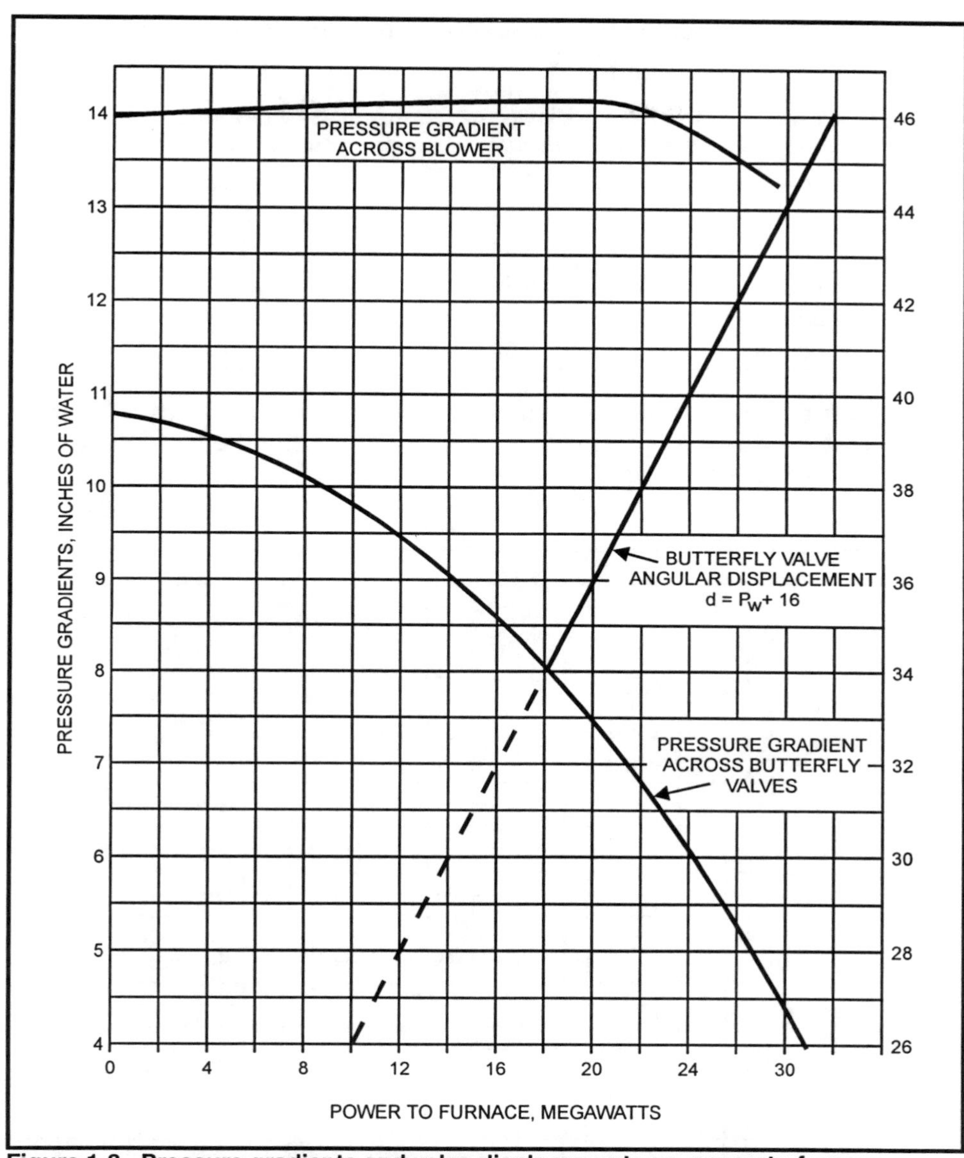

Figure 1-3. Pressure gradients and valve displacements vs. power to furnace

Table 1-2. Exhauster Performance During Periods of Power Change

Open Loop Tests			Closed Loop Tests		
Power, megawatts	Flow Rate, cu. ft./sec	Blower, ΔP in. water	Power, megawatts	Flow Rate, cu. ft./sec	Blower Pressure, in. water
4.0	14.1	7.5	4.5	15.3	8.3
10.0	28.3	7.15	9.0	25.9	8.85
17.0	44.9	6.55	17.1	45.1	7.35
20.0	52.0	6.40	19.8	51.6	7.00
22.5	58.9	6.45	22.5	58.9	6.90
28.8	72.8	6.85	27.6	70.0	7.45

Figure 1-4 indicates that, as power decreases from about 30 to 15 megawatts, the pressure gradient developed by the fans passes through a minimum before again increasing.

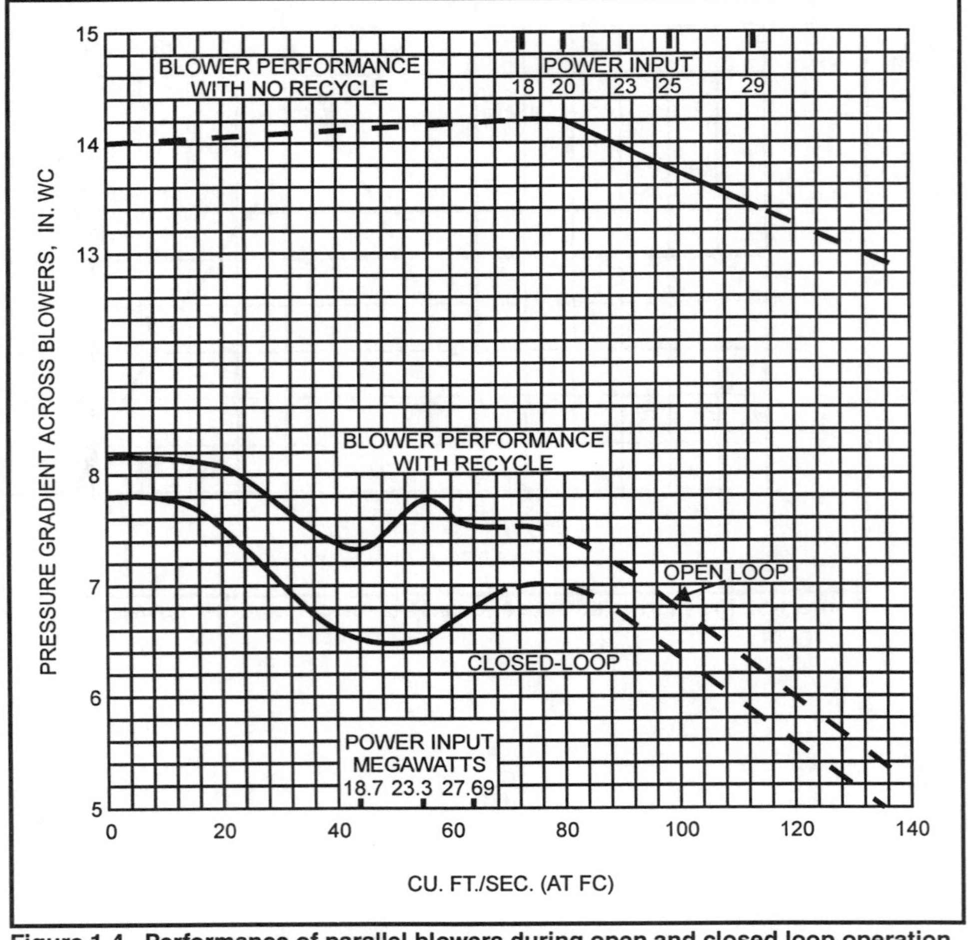

Figure 1-4. Performance of parallel blowers during open and closed loop operation

While evidence is insufficient to provide an unequivocal explanation of this behavior, perhaps differences in the performance characteristics of the two overloaded fans, operating in parallel, may be partly responsible. Nonetheless, this information served to reinforce the recommendation that only a single fan be used in this service, which suggestion plant management eventually accepted.

Some Process Changes

The process was changed by eliminating the recycle and permitting the fans to discharge directly into the low-pressure manifold through a single valve. This arrangement ensured a greater pressure drop across the fan-valve combination, which served to isolate better the system from pressure disturbances appearing at the fan discharge. The valve was provided with a valve positioner to improve its response.

This valve, placed in the combined discharge line from the fans, was sized following the recommendations of the manufacturer and assuming the relationship

$$Q_{60^\circ} = KA\sqrt{g_c\frac{\Delta P}{\rho}}$$

where

Q_{60°	=	flow rate in cu. ft. per second at flowing conditions with the valve 60 degrees open
A	=	superficial cross section of valve, ft2
g_c	=	acceleration of gravity, ft/sec2
ρ	=	density of gas, lb/ft2
ΔP	=	pressure drop across valve, lb/ft2
K	=	an empirical constant taken as 0.885

Valve sizing was also based on the capacity of the valve at different degrees of opening at constant pressure drop relative to the flow rate at 60 degrees open. These capacities are shown in Table 1- 3.

Table 1-3. Relative Capacity of Butterfly Valves at Constant Pressure Drop

% of Flow at 60% Open	Degrees Open
12	20
21	30
34	40
44	45
58	50
77	55
100	60
160	70

The recommended range of operation lies between 30 and 55 degrees open.

From other sources the rate of gaseous effluent from the furnace was taken to be about 3380 $ft.^3/min.$ at 60°F and 2 in. WC when the power consumption was 26 megawatts. The molecular weight was assumed to be 25.8 and a density at 60°F of 0.06 lb/ft^2. A new butterfly valve was obtained designed to handle the indicated quantity of gas at the prescribed conditions. This turned out to be a 24-inch valve which, incidentally, matched the size of the existing piping. Flow rates during the subsequent tests were computed from the valve specifications at existing flow conditions.

Preliminary Dynamic Tests of Process

These tests were performed to observe the dynamic response of the system components. The first test recorded the response of furnace pressure to a change in power to the furnace. The test was executed by briefly interrupting the power supply to the furnace by opening and closing the switches (massive devices immersed in oil). Power was interrupted for about 5 seconds.

Oscillographs from one test are shown in Figure 1-5. Departure of the input power signal from a true rectangular pulse was caused by the dynamics of the current transformer, which had a first-order time constant of about 0.3 seconds. The response of furnace pressure was immediately recognized as being essentially first-order! The time constant was about 3 seconds. A delay time of 0.15 seconds was also present. (Very likely this was the first test of this kind ever executed in the processing industry from which the form of a dynamic relationship could be immediately recognized by merely inspecting the time histories.) Later, similar tests were executed at different power levels and the test data converted to frequency responses using the trapezoidal approximation of the Fourier transform. These results (shown in Figure 1-6) suggest a dependence of response time on the rate of energy absorbed. On the other hand, response may be largely dependent upon both the volume of free space in the furnace and the resistance offered by the transfer line connecting furnace and dust separator. Neither free space nor line resistance probably remained constant. Aside from the above, these tests served to demonstrate the dependence of the rate of this highly endothermic reaction upon energy input.

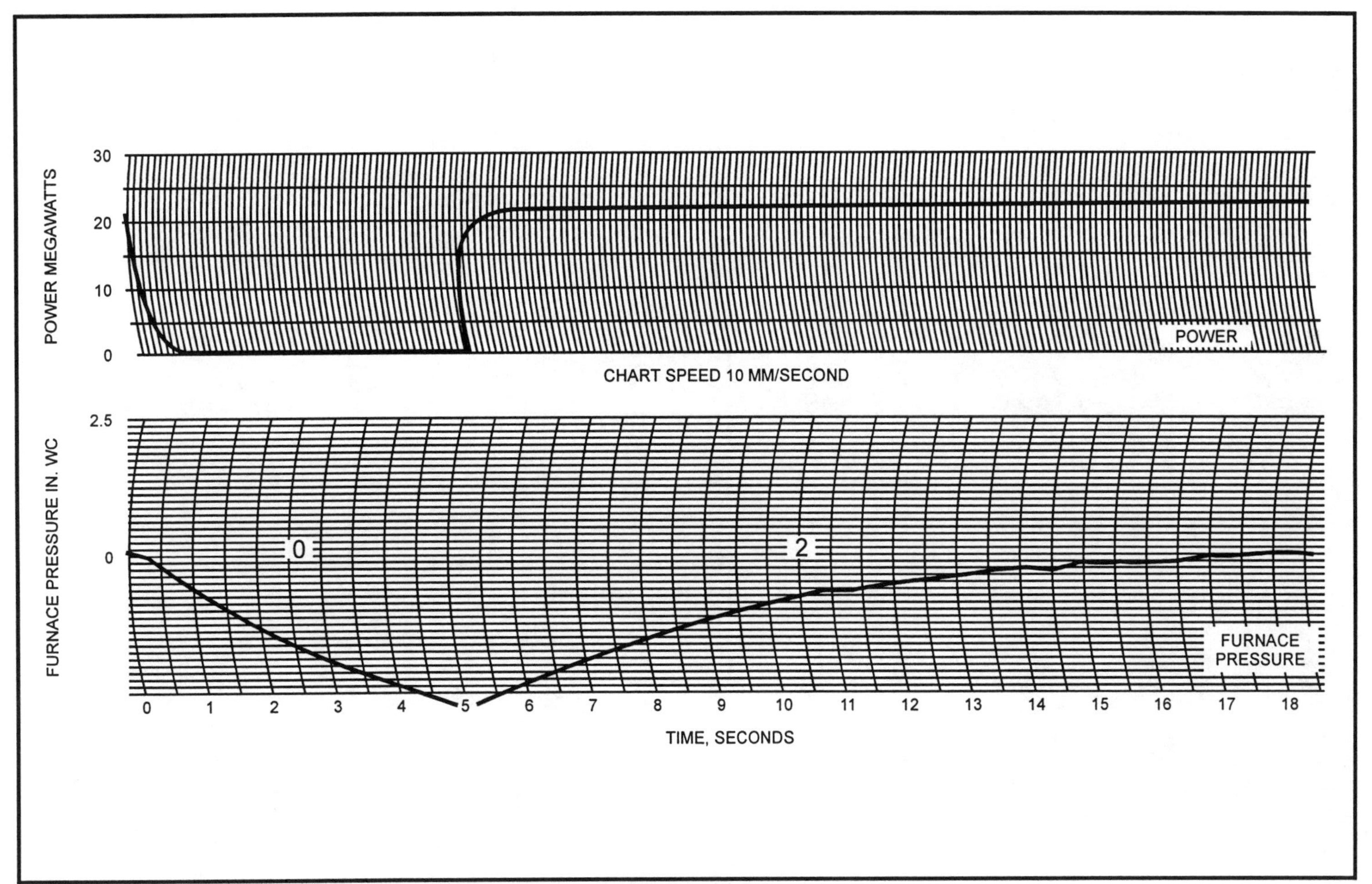

Figure 1-5. Response of furnace pressure to an abrupt reduction in power input

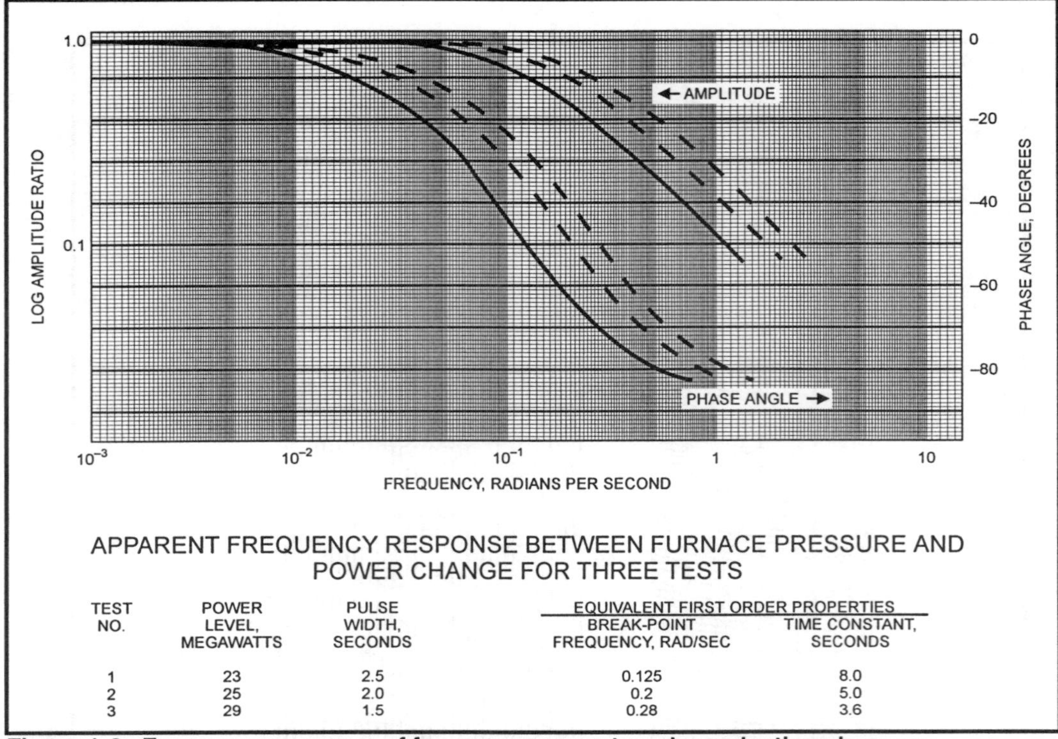

APPARENT FREQUENCY RESPONSE BETWEEN FURNACE PRESSURE AND POWER CHANGE FOR THREE TESTS

TEST NO.	POWER LEVEL, MEGAWATTS	PULSE WIDTH, SECONDS	EQUIVALENT FIRST ORDER PROPERTIES	
			BREAK-POINT FREQUENCY, RAD/SEC	TIME CONSTANT, SECONDS
1	23	2.5	0.125	8.0
2	25	2.0	0.2	5.0
3	29	1.5	0.28	3.6

Figure 1-6. Frequency response of furnace pressure to pulse reductions in power

Dynamics of the Revised System and Control Components

We were now in a position to obtain information needed to design the pressure control system. With all pertinent control systems in the manual mode and power input at a steady value, the terminal butterfly valve was stroked in a pulse-like manner, and the response of furnace pressure measured. Time histories of the signal to the valve and the response of pressure at the furnace are shown in Figure 1-7, where a delay time of about 0.5 seconds may be observed. The frequency response, also shown in Figure 1-7, demonstrates first-order characteristics, with a break-point frequency of about 0.6 radians per second, *i.e.*, a first-order time constant of 1.7 seconds.

These dynamic properties of the process clearly signified that all components of the control system for this process required care in their selection; in addition, the dynamic response of each must be known. Both the static and dynamic characteristics of all proposed control components therefore were determined experimentally using pulse excitation to obtain dynamics, with time histories reduced to frequency response by the trapezoidal approximation of the Fourier transform computation routine. These results are described later.

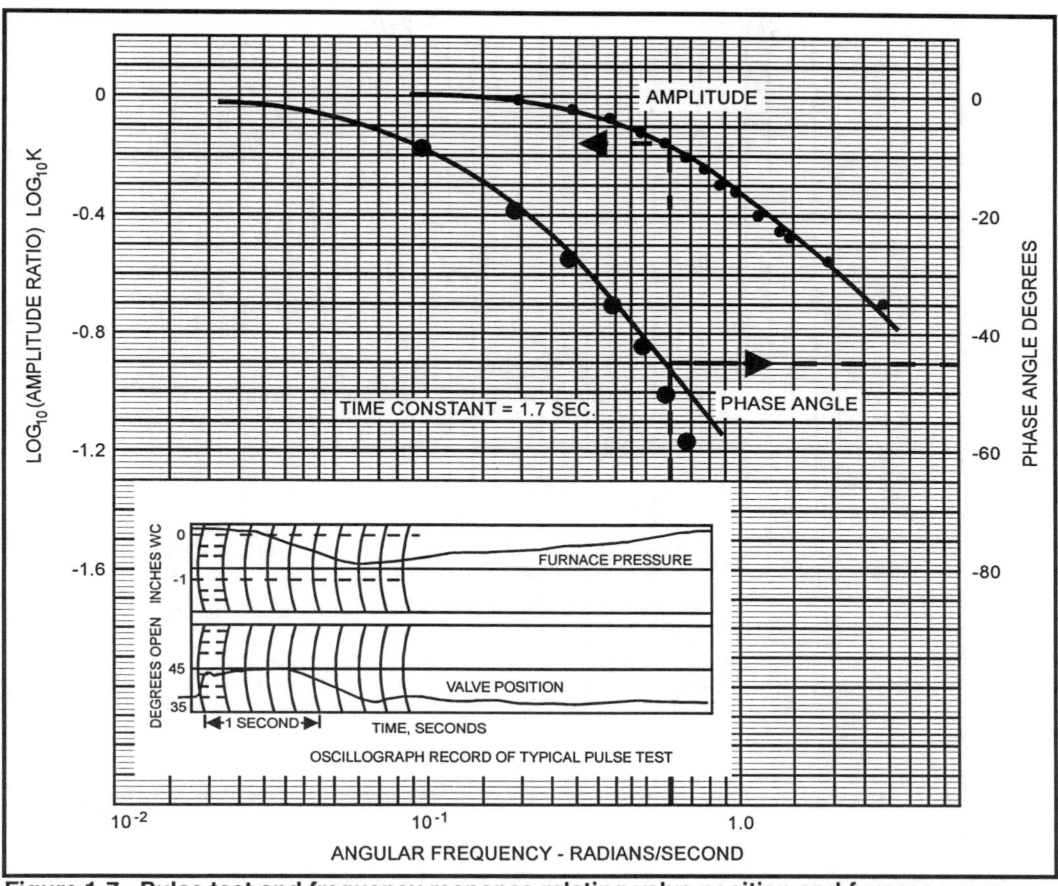

Figure 1-7. Pulse test and frequency response relating valve position and furnace pressure

Developing a Feedforward Control Strategy

Since the rate of energy input was the dominant measurable variable most directly affecting furnace pressure, the feedforward use of a measure of power to position the terminal valve appeared attractive. At different power inputs, the pressure to the valve positioner that would place the valve in the position required to maintain the furnace pressure at 1/4 in. of water was found. The results are presented in Figure 1-8. The linear relations obtained were d = Pwr + 16 and $p_v = 0.18z$ where

z	=	valve displacement or position
Pwr	=	power to furnace
p_v	=	pressure to valve positioner

Combining these gives p_v = 0.18Pwr + 2.84

This strategy was incorporated into the recommended control system, perhaps the first instance in the chemical industry in which a feedforward process control scheme was implemented using experimental data.

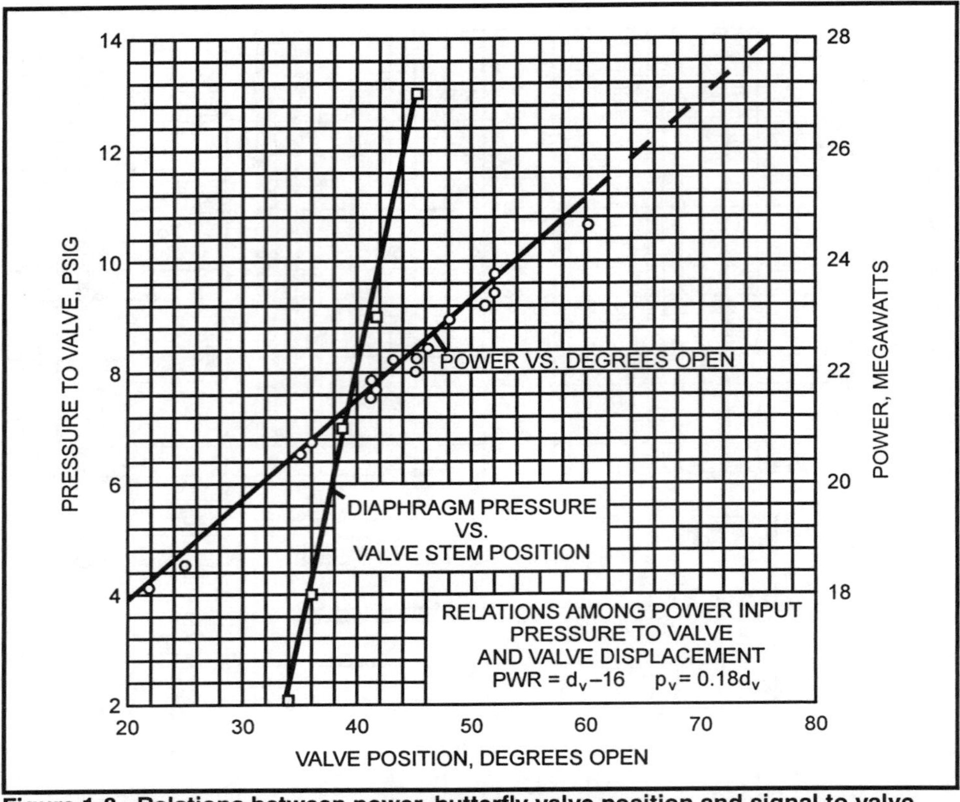

Figure 1-8. Relations between power, butterfly valve position and signal to valve consistent with furnace pressure of 1/4 in. water

Completing the Control System

Figure 1-9 is a block diagram of the proposed control system in which both the feedforward power signal and an air pressure signal of 2.84 psi are added to the output of the pressure controller. The purpose of the dynamic functions of the controller is to compensate for the dynamics of all the remaining components in the feedforward path. The dynamics of each of these components were obtained experimentally and are shown in each block in Figure 1-9. The pressure sensor, positioner, valve actuator, and furnace each possessed first-order properties, but the relay possessed lead characteristics, compensating in part for the lag in the pressure sensor. With two first-order lag elements in the feedforward path, a controller having two compensating lead elements was considered desirable. Towards this end a standard three-mode pneumatic controller was modified to include an additional "derivative" function. When placed in service the ensemble operated superbly. However, in the course of experimental adjustments of the controller parameters, the discovery was made that the extra "derivative" component in the controller could be deleted without adversely affecting control, thus, it was removed to simplify its use and maintenance.

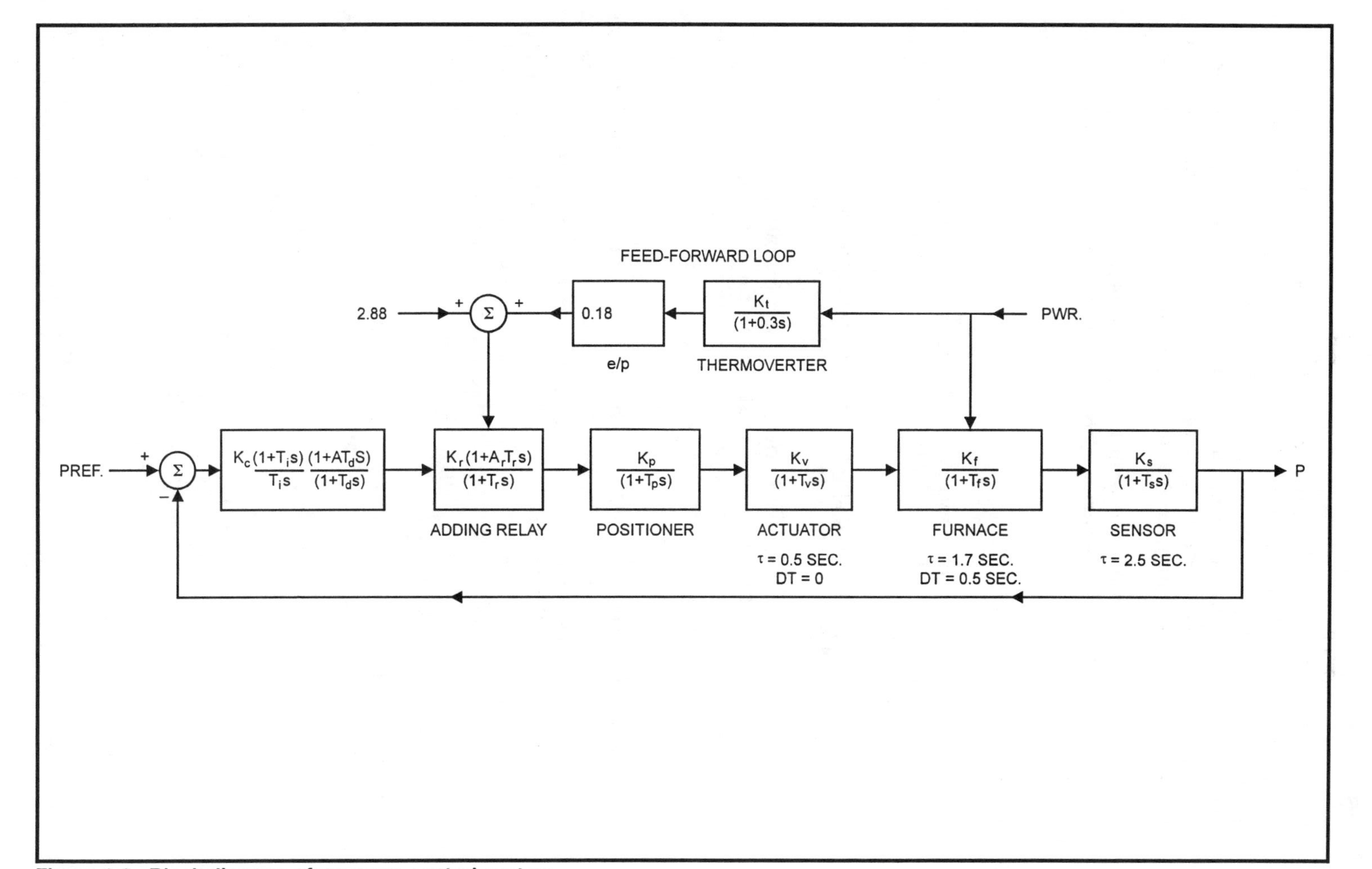

Figure 1-9. Block diagram of pressure control system

The Final Result

The records in Figure 1-10 show how well the final control system responded to both moderate and abrupt changes in power occurring over a period of 48 minutes. The maximum deviation of furnace pressure never exceeded 3/4 in. WC even on occasions when power diminutions were very abrupt. The performance of the valve was especially gratifying. The very steady pressure gradient across the fans should also be noted.

A remark by a veteran plant operator was most rewarding: "This is the best work done to improve furnace pressure control since this furnace was installed."

What Was Learned

- Without appropriate and detailed information about a system neither a strategy for its control nor design of the control system can be implemented.
- Sometimes processing apparatus must be replaced or processes rearranged in order to obtain a controllable system.
- Sensors must be capable of detecting, and following faithfully, rapidly changing signals associated with the variables to be controlled. As a rule of thumb the major time constant of a sensor should be 1/10th, or less, of the time constant of the effect it is intended to measure.
- Existing processes may always have rather simple overall dynamics that can be approximated by simple linear forms. One omnipresent and complicating characteristic, however small, is pure delay time. In this study the pure delay time between valve motion and change in furnace pressure (0.5 sec), although quite large compared to the time constant (1.7 sec.) appeared to cause no great difficulty, although even better control would have been achieved had less delay time been present.
- The controller algorithms required for most chemical processes may always be relatively simple.
- If reliable and appropriate experimental data are on hand, simple feedforward control strategies can sometimes be implemented.
- Controller and processing components should be used to attenuate disturbances. In this study, by allowing the fans to operate at a higher discharge pressure and higher pressure differential (no recycle), disturbances arising from the manifolds were reduced.
- Experimental studies, using appropriate data acquisition facilities, will always yield rewarding results for those willing to devote the time and effort required to discover the details.

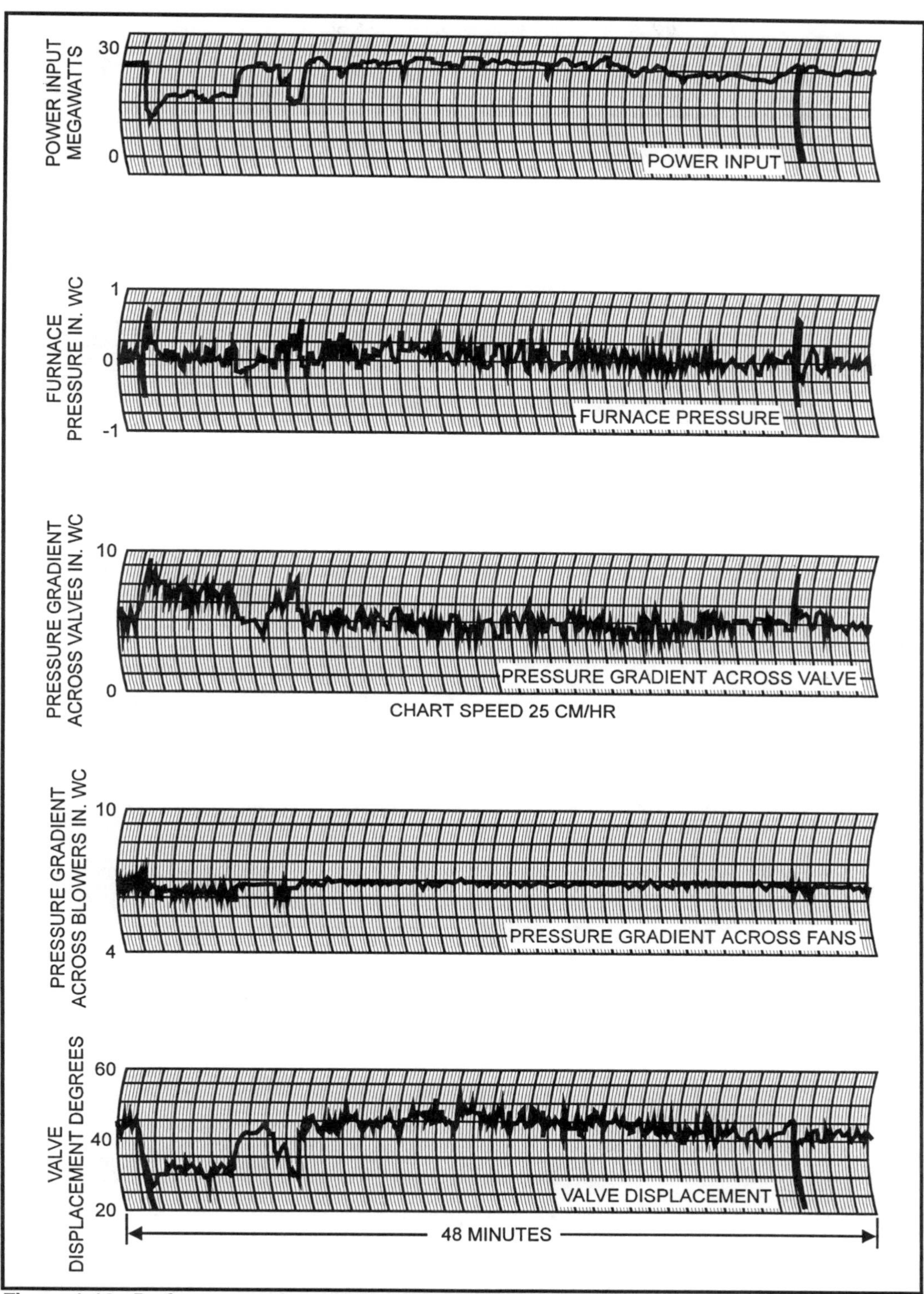

Figure 1-10. Performance of modified system

2

Separation of Alkyl Benzenes by Fractional Distillation

Distillation is one of the most common of all "unit operations" found in the processing industries. Perhaps no other single process is more widely used, studied, and described in the literature or figures more prominently in the economics of chemical production.

Why then do we have the temerity to add to already profuse literature? The reason: we believe this case study, as with all studies included herein, discloses some information, techniques, and insights not found in other sources, and may be helpful to others faced with processing problems in a variety of industries. This example of plant testing was chosen because the objectives and difficulties are typical of those likely to be encountered in many distillation processes.

Definition of Distillation

Fractional distillation may be defined as the process by which a vapor stream, rich in volatile components, is contacted in a counter-current manner with a liquid stream, richer in less volatile components, so that the vapor stream is enriched in and the liquid stream deprived of more volatile components.

The vapor stream is produced by supplying thermal energy to less volatile material in the bottom of the device, usually a vertical cylindrical vessel containing a number of trays, or plates, provided with risers through which the vapor ascends and down-comers through which liquid descends. Thermal energy for vaporization enters through a heat exchanger or reboiler located at the bottom of the column. It is here that bottoms material is partially vaporized and continuously enters the lower end of the column. The liquid is formed by condensing overhead material, a portion of this lighter material is returned to the top of the column as reflux. As vaporized material rises and reflux descends, the desired exchange of constituents occurs.

The processes proceed because of influx of thermal energy at the bottom and extraction of energy from the top. In fact, upward of 80% of energy injected through the reboiler is usually removed though the overhead condenser.

Although steady-state performance is the principal focus, the observed transient behavior of the column and system components is mentioned. This study, as with most presented herein, was performed before the advent of computers and modern instrumentation. Nonetheless, the lessons learned and information retrieved are of value today.

The Purpose of the Study

The specific aim of this study was to improve the performance of distillation facilities used to separate a mixture of alkylated benzenes in order to obtain a bottoms product rich in C_{11} and C_{12} and as free as possible from components characterized as C_{10} and lighter. The desired product was used to make detergents before those containing phosphates were considered ecologically undesirable.

The Process

A mixture of hydrocarbons, including components from C_8 to C_{13} containing the most desirable, C_{11} and C_{12}, was to be separated into an overhead and bottoms, the latter containing most of the C_{11} and C_{12}, with a minimum of C_{10} and lighter, more volatile, or lower molecular weight components. The separation was to be accomplished using a distillation column about 5 feet in diameter with 20 bubble cap trays spaced about 24 inches apart. The vacuum column was operated at about 50 mm Hg top pressure using a steam jet with controlled air injection to regulate this sub-atmospheric pressure. Feed was introduced on tray no. 6.

Reboiler duty was supplied by steam, the flow rate of which was regulated by a flow controller, which in turn was adjusted in accordance with the signal from a thermocouple located in the vapor space above tray no. 18, near the top of the column. A diagram of the original facilities, including the various control systems, location of sensors, valves and other items is shown in Figure 2-1.

Attention is directed to the method of regulating the temperature near the top of the column. The sensor of this temperature was a thermocouple inserted into a heavy thermal well located in the vapor space above tray 18. The signal from this sensor was compared to a reference temperature (setpoint), the difference directed to a temperature controller. The output of this controller formed the reference to, or setpoint of, the cascaded flow controller regulating the flow rate of reboiler steam. Other features of the original arrangement are pointed out later.

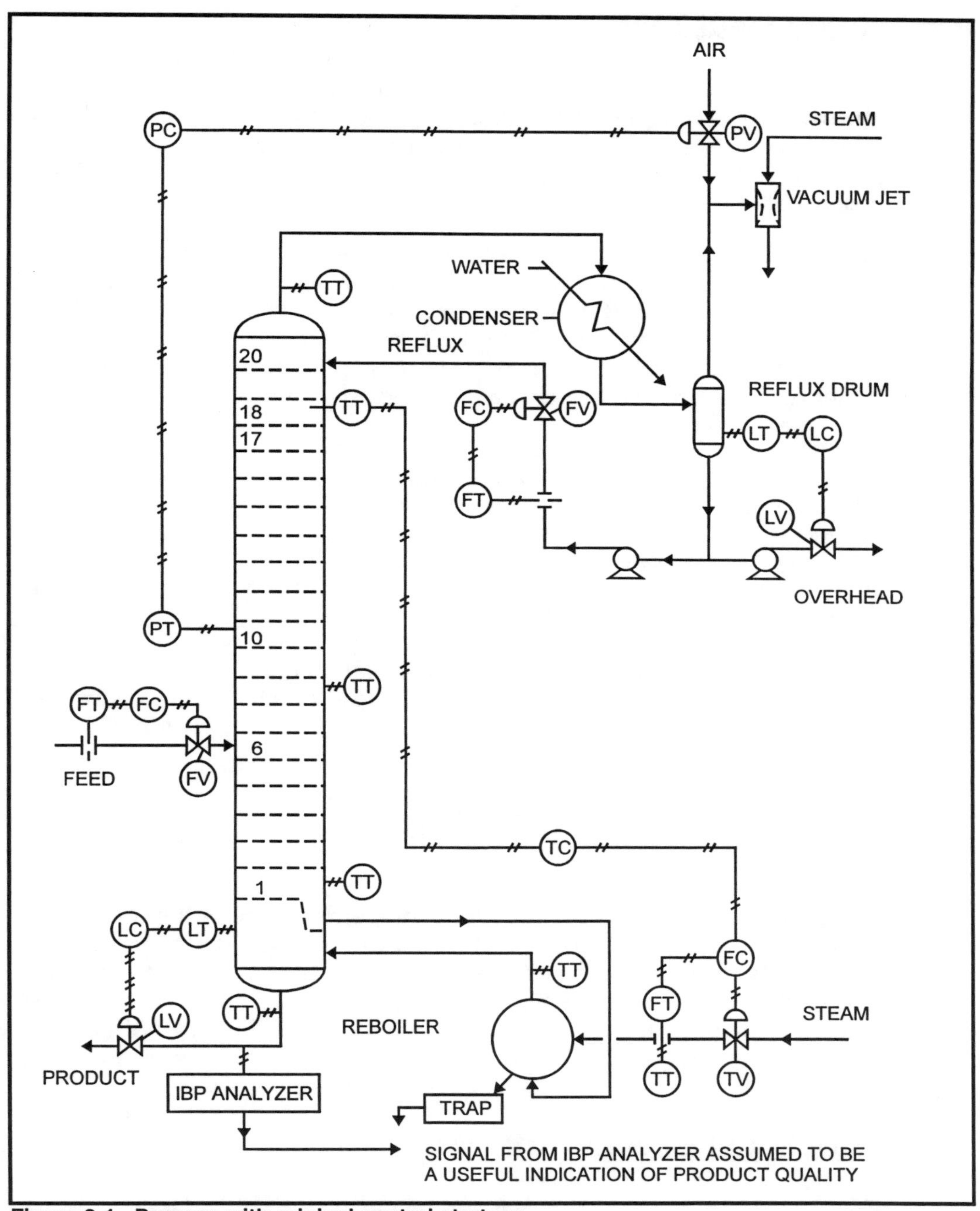

Figure 2-1. Process with original control strategy

Preliminary Field Tests

Preliminary tests, with certain controllers in the manual mode, immediately revealed that dynamic response of the temperature sensor located above tray 18 was exceedingly slow. In addition, dynamic response of the initial boiling point (IBP) analyzer was suspected as being inadequate for the purpose intended.

The generally slow response of temperature sensors, particularly if located in a vapor environment, is caused by the resistance to heat transfer offered by the fluid film around the enclosure containing the sensing element. The sensor may be either a thermocouple or resistance element, which responds to temperature. Industrial sensors are usually inserted into rather massive protective wells, which also adversely affects transient response.

The dynamic response of the initial boiling point instrument was tested by introducing, in a step-wise manner, changes in the composition of feed to it. Feed with differences in both light and heavy components were introduced in separate tests. Time histories of the tests of the IBP analyzer (not shown here) indicated that its response was essentially first-order with a time constant of 1-1/2 minutes and with a pure delay time of approximately 3 minutes. In addition, a regression analysis, employing 50 samples of bottom product, showed a correlation between IBP and product composition having a reliability of less than 50%. This instrument was obviously entirely unacceptable for indicating composition or as a component in a feedback control loop.

Standard ASTM distillation test data, such as IBP, 1, 2, 3, or 5 ml(%) points, or slopes of ASTM distillation tests, also proved of little value in characterizing the bottoms product.

Some tests were also made to evaluate the performance of the control systems, in particular the temperature controller associated with tray no. 18. Records were obtained of several variables following abrupt changes in the set-point of the top temperature controller, as well as disturbances entered elsewhere. In all cases, responses were very sluggish with temperatures and pressures attaining desired values only after inordinate lengths of time.

Results of Initial Experimental Work

Preliminary findings are summarized below:

- The IBP analyzer was not suitable for this application.
- All measurements of process variables were suspect, and the response of temperature sensors was especially slow and uncertain.
- Some hysteresis was present in all control valves.

- Changes in column pressure were caused both by poor control of the vacuum on the column and by variations in the flow rate of steam to the reboiler.

A detailed study of the system seemed required, as well as more information about the fluid being processed.

Analysis of Feed Stock

A separation was first obtained of the more volatile 30% by weight of the feed stock using an Oldershaw fractionating column having about 25 equivalent trays operating at a vacuum pressure of 10 mm and a reflux ratio of 40 to 1. The first 60% by volume of this primary fraction was then separated in a Podbielniak Mini-Cal® distillation column into 12 successive fractions, each equal to about 1% by volume of the original feed. Specific gravity and refractive indices were measured, and mass spectrometer analyses were performed on each of these fractions.

The following observations were forthcoming from the laboratory fractionations:

- A rather abrupt break in the true boiling point curve obtained from the Oldershaw distillation occurred at about 17% by weight distilled, at which point the temperature was 142°C. at 10 mm Hg vacuum. This point appeared to coincide with the appearance of a preponderance of C_{12} in the distillate.
- While material lighter than C_9 could be separated fairly well in the Podbielniak Mini-Cal® column, C_{10} did not separate sharply, and C_{11} and C_{12} were almost identical in volatility in the fractions in which they appeared.
- Refractive index and specific gravity both showed similar dependence upon composition and exhibited abrupt changes in the vicinity of C_{12}. An almost linear relationship between refractive index and specific gravity exists, based on data from the discrete fractions obtained this study.
- Compared to the separation achieved by the Podbielniak Mini-Cal® column, the plant was achieving poor separation. For this reason, the use of either specific gravity or refractive index as a monitor of column performance or product quality did not appear feasible. The separation achieved by the Mini-Cal® column is shown in Table 2-1.

Figure 2-2, constructed from the laboratory distillation data given in Table 2-1, clearly emphasized the problem of separating the various components in a typical feed stock. As more overhead is produced, C_9 and C_{10} are completely removed; but as this occurs, both C_{11} and C_{12} continue to appear in the distillate. Obviously C_{11} and C_{12} are virtually inseparable by fractional distillation and are removed in an almost linear fashion in the very efficient laboratory apparatus.

Table 2-1. Separation Achieved by Laboratory Distillation

Volume % of Distilled Product	Cumulative Volumetric % in Distillate			
	C_9	C_{10}	C_{11}	C_{12}
9	6.0	0.2	0	0
10	10.2	3.4	1.2	0.4
11	9.6	4.4	5.1	3.6
12	8.9	4.4	8.0	7.4
13	8.2	4.2	10.5	11.4

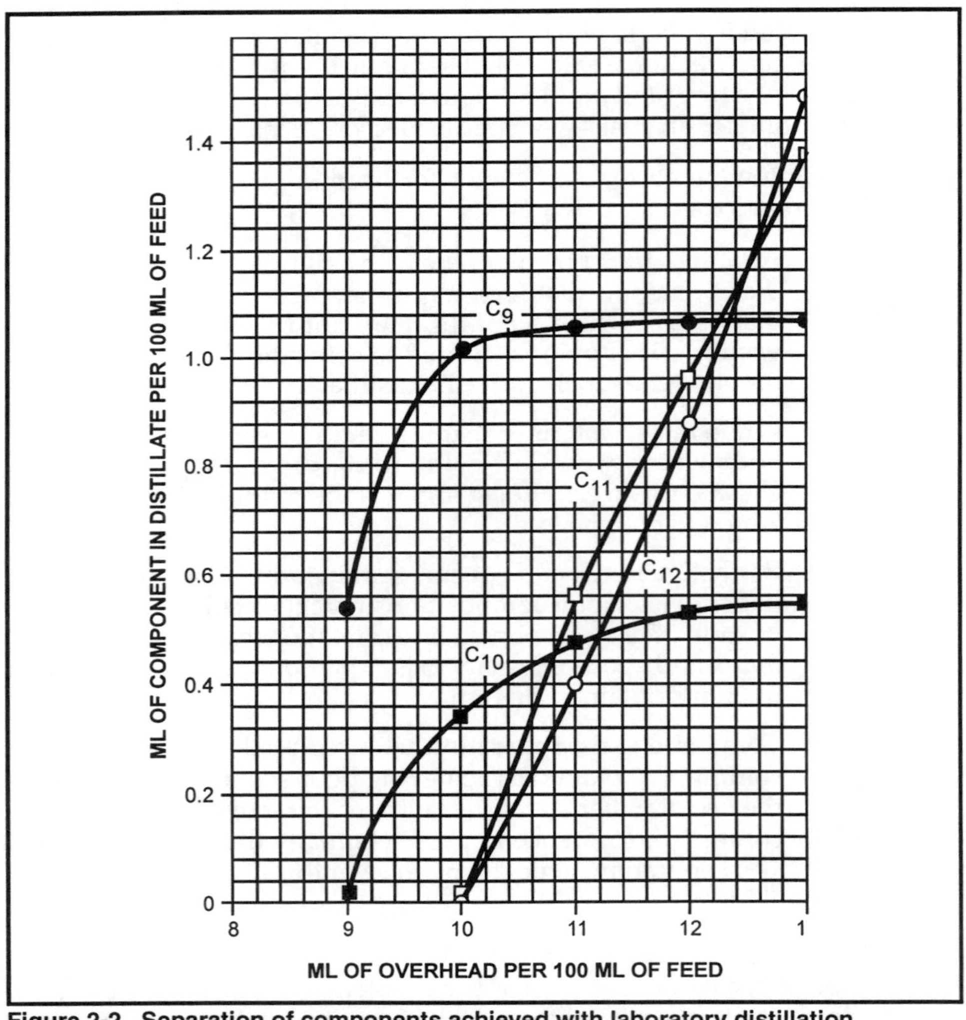

Figure 2-2. Separation of components achieved with laboratory distillation apparatus

The Test Data System

Since plant instrumentation was both inadequate and insufficient for satisfactory data acquisition, special test instrumentation was procured and installed. The data acquisition system consisted of twelve channels of recording oscillographs, each provided with high-gain, high-impedance dc amplifiers. Signals from test transducers were transmitted through shielded conductors with a common ground at the data system, which was located in the plant control room. Across each amplifier input was a set of capacitors, one of which could be selected in order to remove signal corruption that might appear at an arbitrary frequency. Each channel was also provided with signal suppression capability so that any signal could be suppressed a known amount. The suppressed signal could then be amplified as desired; thus deviations as small as 2 to 3 microvolts could be recorded on the assigned oscillograph channel. Chart speed could be changed to meet the needs of the particular test.

Temperatures of greatest interest were measured with resistance elements, pressures and pressure differentials with strain gage transducers, and valve stem displacements with rotary potentiometers. All items were calibrated prior to installation. The location of test sensors is shown in Figure 2-3.

Observations of Process Performance

Behavior of the column was observed at various steady state conditions; in addition, a number of cursory dynamic tests were executed. During many of these tests, samples were collected, especially of the bottoms product. These samples were subsequently analyzed by mass spectrometer. During these tests, column pressure, column bottom, and reflux accumulator levels were under control by their respective regulators. Otherwise the open-loop mode, or manual control, was exercised insofar as was practicable. Thus, feed flow rate, reboiler steam valve stem position, and reflux flow rate were under manual control during these testing periods.

During those tests in which the steam supply valve was maintained in a fixed position, the flow rate of steam to the reboiler unavoidably varied. This occurred because steam flow rate was a function of both valve stem position and pressure drop across the valve. The latter could, and did, change during tests by virtue of changes induced in the pressure of steam within the reboiler. (This observation has important implications that are mentioned later.)

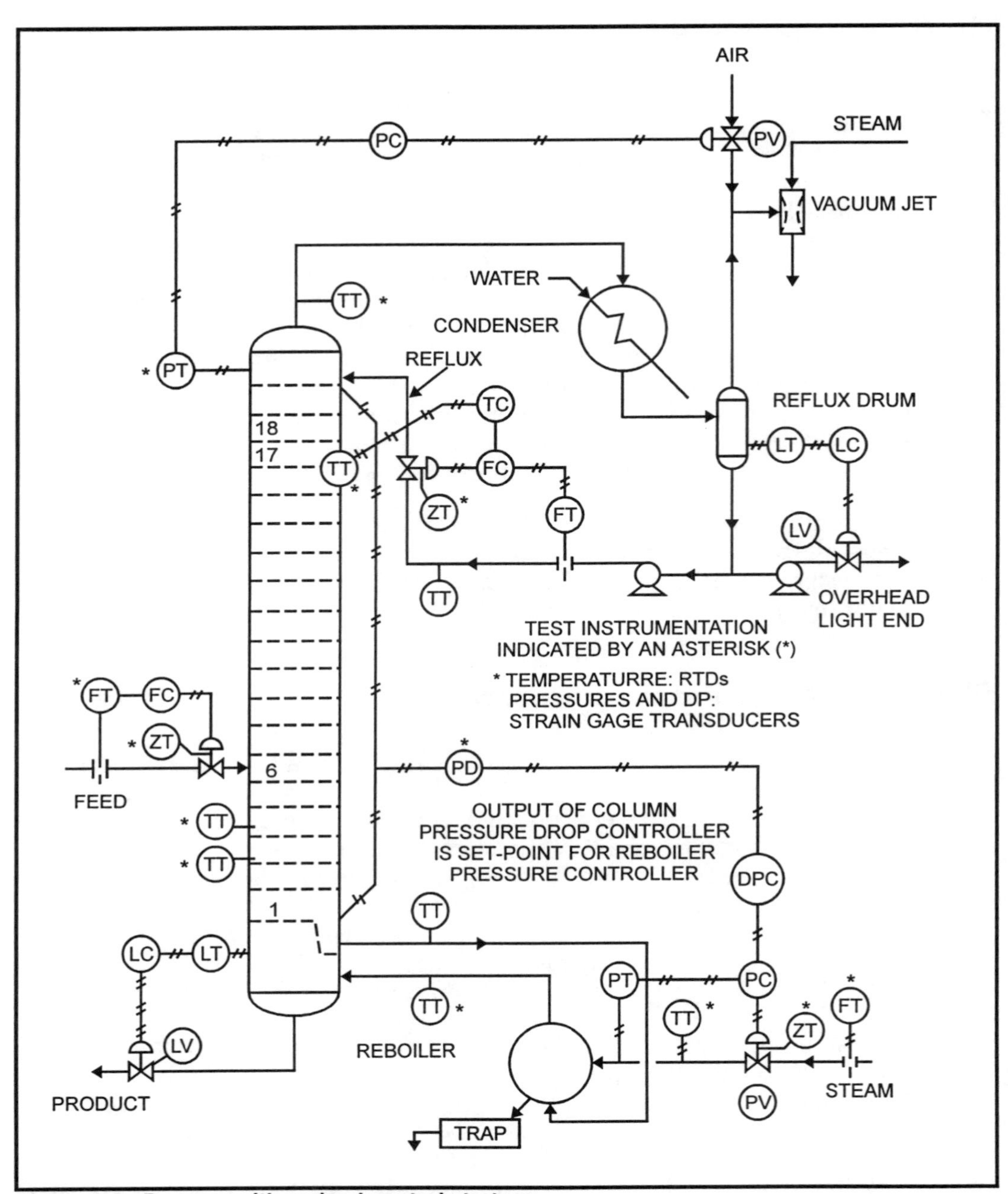

Figure 2-3. Process with revised control strategy

Results of Plant Tests

Some of the more important observations derived from these tests are summarized.

- The well-known interdependence of various phenomena that determine the behavior of a distillation column was quite apparent at all times. The interplay of these interactions was sometimes rather subtle, however.

- Temperatures in the bottom section of the column were relatively slow to respond. Reboiler inlet and outlet temperatures would respond to changes in steam flow rate, but such changes would be slower and smaller compared to those changes induced in the upper portions of the column.

- The pressure drop across the column was highly dependent upon, and responded rapidly to, changes in reboiler duty. The relationship between the two could be approximately described, dynamically, as first-order.

- Changes in feed rate influenced column temperatures only mildly, but changes in feed temperature were quickly detected.

- Changes in reboiler duty produced immediate and relatively large changes in the temperature of the liquid on trays above the feed tray (tray no. 6). The 17th and 18th trays exhibited the largest perturbations.

- Temperatures in the upper portion of the column were more sensitive to changes in reflux flow rate than to any other variable. Again the largest perturbations occurred on trays 17 and 18.

- Following a change in reflux flow rate, a change in column pressure drop occurred. However, such changes seemed to be caused in part by accompanying self-induced changes in the flow rate of steam to the reboiler.

- Since the existing pressure controller operated off the 10th tray, both top and bottom pressures could, and did, change in response to changes in dependent variables.

- Changes in composition of bottom samples appeared to occur not much sooner than two minutes following any change imposed on the column.

Conclusions

From the above, and other information, the conclusion was reached that the column could be operated in a satisfactory manner if the following variables were controlled:

- Pressure at the top of the column
- Temperature on Tray 17
- Pressure drop across the column

Control of top pressure ensured the significance of the temperature on tray 17, and, as had been demonstrated, the response of this temperature to disturbances was about the maximum and most rapid of any column temperature. Finally, since the value of the product was high, operating the column most efficiently was desired, *i.e.*, near its flooding point, which, in turn, could be detected by pressure drop measurements. In addition, with both column top and pressure gradient controlled the temperature near the bottom of the column would also be more meaningful.

Results Following Recommended Changes

After the changes were made, subsequent operations were very satisfactory, not only from an operational point of view but from the increased profits obtained. Not only was the top temperature well controlled but also the pressure differential across the column. Temperatures in the lower part of the column now served well as indications of the composition of the bottoms product.

Among the original objectives was the determination of the transient response of various dependent variables to changes in selected independent variables. The purpose was to obtain information needed to design the control systems. However, by merely experimentally adjusting the parameters of the controllers, the performance of the process, with the revised control arrangement, was so much improved that those studies were postponed.

Some Additional Tests with Surprising Results

Among the more interesting dynamic tests conducted on this process was the response of the column to changes in the composition of the feed. An arrangement was made by which either overhead or bottoms material could be introduced into the feed stream in a pulse-like manner, while maintaining the feed flow rate constant. For this purpose a small vessel was connected by a pipe to a point upstream of the feed-flow meter orifice plate. Installed in this pipe was a calibrated turbine flow meter and a small pump.

In the first test some of the overhead material, from the column reflux drum, was transferred to the special test vessel. In the interim the material had time to cool substantially. Then, at a prescribed moment, material from the test vessel was pumped, at a measured and constant flow rate equal to about 10% of the feed rate, into the feed stream for a brief period. The results were entirely unexpected!

Instead of a gradual change in overhead production and in other variables, the level in the reflux drum very suddenly increased until the added material was removed. Changes in temperatures below the feed tray were almost imperceptible!

In a like manner bottoms material was injected into the feed stream. The results were analogous to the above: the heavy material simply descended to the bottom, while conditions above the feed tray remained virtually unchanged.

The light material injected did not simply flash because it was well below its boiling point at column pressure. Yet it, or material of corresponding volume, remained in the column only very briefly. Although the material appeared to vaporize quickly, temperatures below the feed tray changed hardly at all; the sensitive temperature sensors located a few plates from the bottom showed little response. It is less surprising that the heavy material injected should rather quickly appear in the bottoms, but one would expect some small change above the feed tray. These were not observed.

Estimated Economic Benefits from Recommended Changes

An analysis of the fractions obtained from the Mini-Cal distillations gave the results previously shown in Table 2-1. From these data Figure 2-2 was prepared, from which it is apparent that, following the distillation of about 11% of the feed in the laboratory column, virtually all the C_9 and C_{10} material has been removed. Also the almost identical volatility of C_{11} and C_{12} fractions is evident.

Assuming the plant column, provided with the revised control system and operating near the flooding point, could separate the feed so that various fractions of the desired material shown in Table 2-1 could be recovered, the potential economic benefit was computed. These results are shown in Table 2-2 and are based on a plant on-stream time of 80%, an average feed rate of 100 gal/min, and a selling price of 10 cents per pound.

Table 2-2. Estimated Benefits Derived from Modified Separation System

Fraction of Maximum Potential Recovered	Increase in Product		Savings, $/year
	gal/min	lb/year	
1/8	0.25	709,100	74,070
1/4	0.50	1,418,200	149,050
1/2	1.00	2,836,460	298,100

What Was Learned from This Study?

- The performance of existing instrumentation should always be suspect. Sensors may be installed incorrectly, in the wrong environment or enclosed in protective sheaths that impede their response, may be poorly calibrated, and may lack sensitivity. For example, in this study the existing thermocouple above tray 18 was inserted into a heavy thermowell and, moreover, was located in the vapor above that tray. It is well to bear in mind, also, that the output from this thermocouple was equivalent to about 30 microvolts per degree C. Thus, to observe changes less than one degree would have required a very sensitive detecting system. The resistance element that was substituted for a thermocouple was arranged to produce an output equivalent to 1 millivolt per degree, 30 times larger than that from the thermocouple! (The test data system could measure signals as small as 2 microvolts.) It should also be borne in mind that thermocouples made from random selections of wires from two spools may have outputs differing as much as 4°F. from the published tables. This could spell disaster when measuring small temperature differences. Finally, the performance of analyzers should always be verified, both as to accuracy and transient response.

- Control valves usually have hysteresis; for the valves tested in this study, this was not particularly serious. However, it is good practice, when testing processes, to measure valve stem position, signal to the valve positioner, and air pressure on the diaphragm, as well as the pressure drop across the valve, and, if possible, the flow rate through it. The reasons for procuring such data are not only to verify the valve performance but frequently to use the valve as a reliable meter of flow rate. Some examples are given in subsequent chapters.

- Sensors used for control should be placed in the environment where the change can be expected to be the greatest and most responsive to a given disturbance. In this study the temperature sensor, which was moved from the vapor space above tray 18 to the liquid on tray 17, is an example.

- The control strategy finally developed should ensure that the controlled system will operate as efficiently as possible, especially where improved economics is paramount, thus resulting in the decision to use column pressure drop control in this instance.

- Unexpected behavior sometimes occurs. Light material added to the feed quickly appearing in the overhead, or similarly, heavy material added to the feed appearing in the bottoms of the fractionator, are examples.

- The orifice plate used to sense the flow rate of steam to the reboiler was originally downstream from the control valve. This is not good practice since the properties of the steam, especially pressure, may be subject to change and may differ from that assumed in developing meter factors. It is better to place the orifice plate upstream from the control valve. Also the difficulties of measuring flow rates of steam with orifice plates are well documented and should always be borne in mind. Nozzles and Venturi tubes are preferred for this service.

- Referring to Figure 2-3, which shows the revised control scheme, notice the arrangement of the system controlling admission of steam to the reboiler. Instead of controlling the flow rate of steam to the reboiler, it is the pressure of steam within the reboiler that is controlled. The setpoint, or reference, of the pressure controller, in this instance, is adjusted in response to the signal from the column pressure gradient controller. This was recommended because the pressure of steam in the reboiler determines the rate of heat transfer, not necessarily the flow rate of steam to it. Under some conditions the valve could open, calling for an increase in flow rate of steam, but the downstream pressure may be so high that further increase cannot occur.

- The results of detailed studies leading to improved process control systems often increases profits.

3

High-Pressure Polyethylene Processing

The description of "the problem" was, as is usually true, vague. There were indications that some parts of the process were subject to oscillations, the sources of which were unknown. And, as is so frequently the case, no reliable quantitative information was available from existing instrumentation, operators, or engineers associated with the plant; no satisfactory data were available.

In such cases after a close inspection of the process one must determine, as best one can, the measurements that should be made and with what instrumentation. Ignorance of the true nature of the problem necessitates measuring all variables that might have bearing on assumed or obvious difficulties.

This system was one line of the facility used to produce polyethylene, part of a large complex. The huge reciprocating compressors, producing a pressure of 35, 000 psi, were awesome and could not but instill great respect for mechanical devices and the ingenuity of man.

After studying process and instrumentation diagrams, consulting with operators, mechanics and engineers, and inspecting the plant, 24 measurements were considered sufficient. (The locations of measurements are indicated on the process diagram shown as Figure 3-1.)

The Process

Fresh ethylene, available at about 350 psi and 100°F, is admitted under pressure control to the suction of two "make-up" compressors operating in parallel. Recycled material also enters these compressors, this being the discharge of the "booster" compressor. Recycled material is the light end portion of the final reactor effluent. "Booster" compressor discharge, minus a small purge withdrawal, along with fresh ethylene comprises the

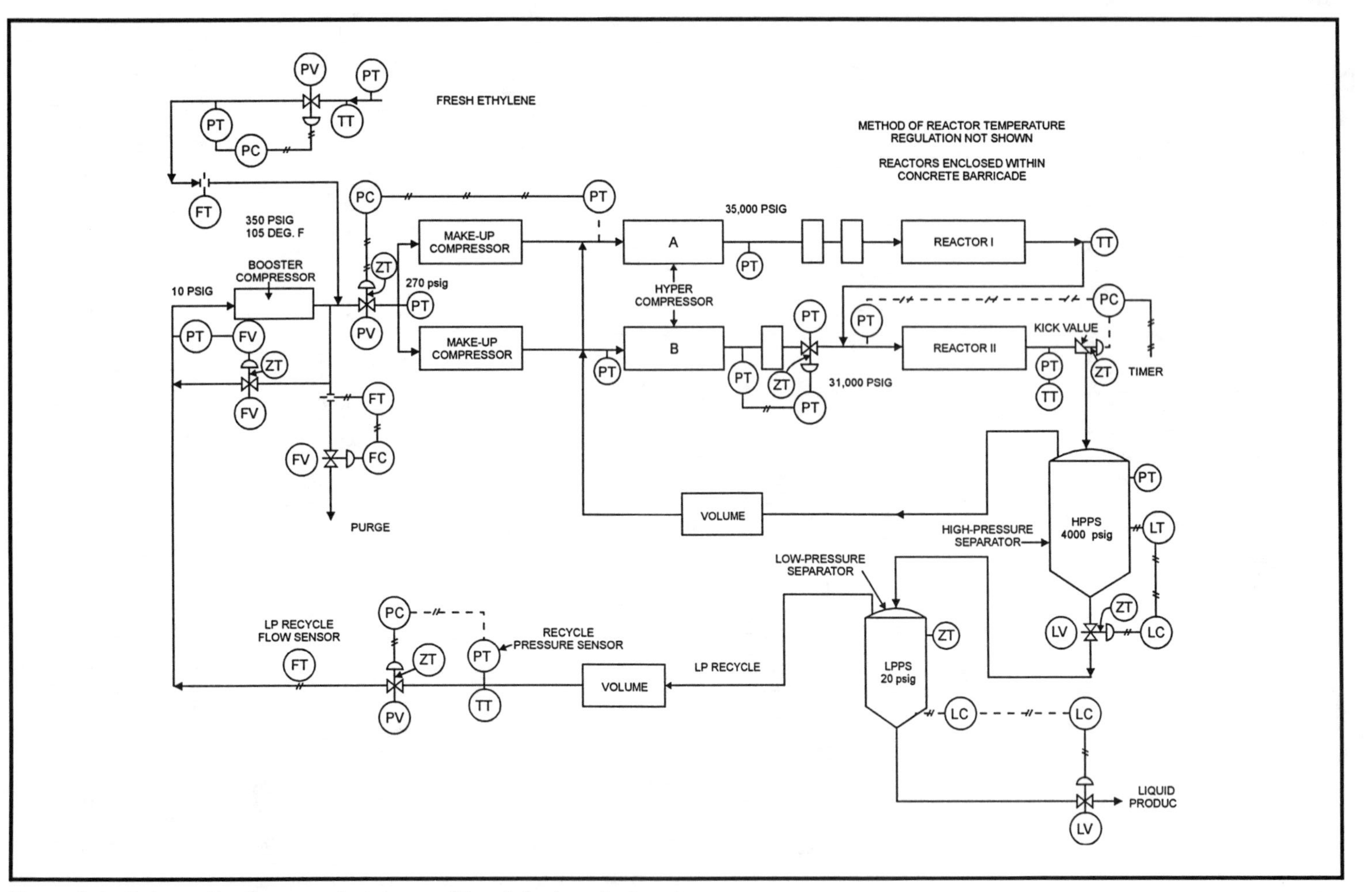

Figure 3-1. Schematic diagram of process with original control system

total input to the "make-up" compressors. Discharge from the "make-up" compressors, at a pressure of 3,500 psi, supplies the parallel "hyper" compressors. Effluent from hyper compressor A (at 35,000 psi) after passing through reactor I, joins discharge from hyper compressor B (at 31,000 psi) the combined stream forming feed to reactor II.

Reactor II discharges into the high-pressure product separator (HPPS) maintained at about 4,000 psi. Light material from that vessel is returned to the hyper compressor suctions, while the heavy phase, or "liquid," flows under level control, to the low-pressure product separator (LPPS) operating at about 20 psi. Liquid product is removed from the LPPS under level control; the light phase is recycled to the booster compressor suction. Various traps and knock-out vessels, which add to the system volume, are provided to ensure liquid-free compressor inlet material.

For a given conversion, gaseous inventory in the two separators is largely determined by the difference between fresh ethylene added and purge emitted. Since the latter is usually fixed at some predetermined level, the inventory, at constant vessel pressures, is dependent upon the admission of fresh ethylene. Residence time within the reactors is regulated to a large extent by the frequency of the opening and closing of a valve located in the discharge of reactor II. In the vernacular of the trade, this valve is commonly designated as the "kick" valve. Abrupt increases in flow rate, accompanying each "kick", obviously produce pressure changes at the suction of the hyper compressors. In addition, flow rate to the make-up compressors is also altered, either by changing booster compressor discharge pressure or by inducing transient variations in the rate of ethylene addition rate. It is beyond the scope of this study to discuss the origin of, and logic behind, this common method of operating these polyethylene-producing reactors.

Changes in flow rate of recycled vapor from the HPPS produce changes in flow rate of material to the "make-up" compressors through hyper compressor inlet pressure control. In addition, changes in flow rate of HPPS liquid, or dense phase, to the LPPS alter the flow rate of effluent gas from the LPPS, and this, in turn, eventually alters hyper compressor suction pressure. A highly interacting system can thus be visualized.

The throughput of the hyper compressors depends on the density of the material at their suctions, since these are essentially constant volume delivery machines. Of the two variables that determine density, temperature and pressure, the latter was found to be more critical. Control of this pressure was not satisfactory, hence, the controller assigned to this service was frequently placed in the manual mode. Since variations in mass flow rate are accompanied by changes in residence time and temperature distribution within the reactors, product quality becomes partly dependent on hyper compressor suction pressure.

The Data System

The data acquisition system consisted of three 8-channel recording oscillographs, each channel was provided with high-gain, high-impedance dc amplifiers with a wide range of amplification adjustment. Across each input a set of capacitors served as filters of noise, or signal corruption, in selected frequency ranges. Additional circuits enabled suppressing steady-state signals by a known amount. Thus very high gains could be employed yet keep the pens on the assigned chart, enabling measurement of very small changes in signals above suppressed levels. Sensitivity was such that deviations as small as 2 microvolts could be measured with confidence.

Pressures and differential pressures were measured with strain-gage-type transducers excited with 10 volts, dc. The full-scale output of these transducers was 5 millivolts per volt of excitation. Frequency response of the strain-gage transducers was flat to about 100Hz. Temperatures were measured with resistance-type sensors calibrated to give 1 millivolt output per degree F. Linear potentiometers with independent power supplies were used to measure valve stem displacements. All sensors were calibrated prior to installation, each with interconnecting shielded signal cables closely approximating the lengths needed to connect field transducers to the special data acquisition system. Some of these cables were in excess of 400 feet in length. The common ground was located in the plant control room along with the data acquisition system. The location of sensors is indicated in Figure 3-1.

Sequence of Experimental Program

The experimental work can be divided into three parts: (1) steady-state observations of plant performance; (2) dynamic testing of subsystems; and (3) evaluation, testing, and calibration of important controllers.

Analyses of oscillograph records under normal operating conditions are presented first. These records enabled locating sensitive areas and identifying malperforming components and reinforced intuition concerning cause and effect. The most significant observations are discussed in the following section.

The HPPS Level Control System

This control system was found rather quickly to be unsatisfactory. The level signal itself was questionable but, more importantly, when operating in the automatic mode, the "controlled" variable was highly oscillatory. This behavior could be caused by excessive gain in the feed-forward path, valve hysteresis, an incorrect controller algorithm, or possibly all three. In addition, the inlet to the pressure tap to which the level sensor was attached on this vessel may have been exposed to local disturbances arising from the rapid expansion and circulation of the material in this vessel, which may have been undersized.

Oscillatory levels are not always particularly troublesome, but in this case the effects of the HPPS level variations were pervasive, producing variations in HPPS pressure and rate of discharge, as well as changes of pressure in the low-pressure recycle stream. Cyclic pressure changes occurring in the HPPS also appeared in hyper compressor suctions. These oscillations were independent of the effect of the "kick" and had a frequency of about 0.7 cycles per minute. The frequency of the "kick" at the time was about 2.4 per minute.

Figure 3-2 shows oscillograph records of HPPS pressure (top), HPPS level signal (center), and deviations in the position of the level control valve (bottom). The sharp, nearly vertical, spikes in HPPS pressure were caused by the action of the "kick" valve, which periodically released material from the 31,000 psi reactor. These spikes in pressure are approximately constant at 65 psi and are superimposed on cyclic oscillations of pressure of a lesser magnitude.

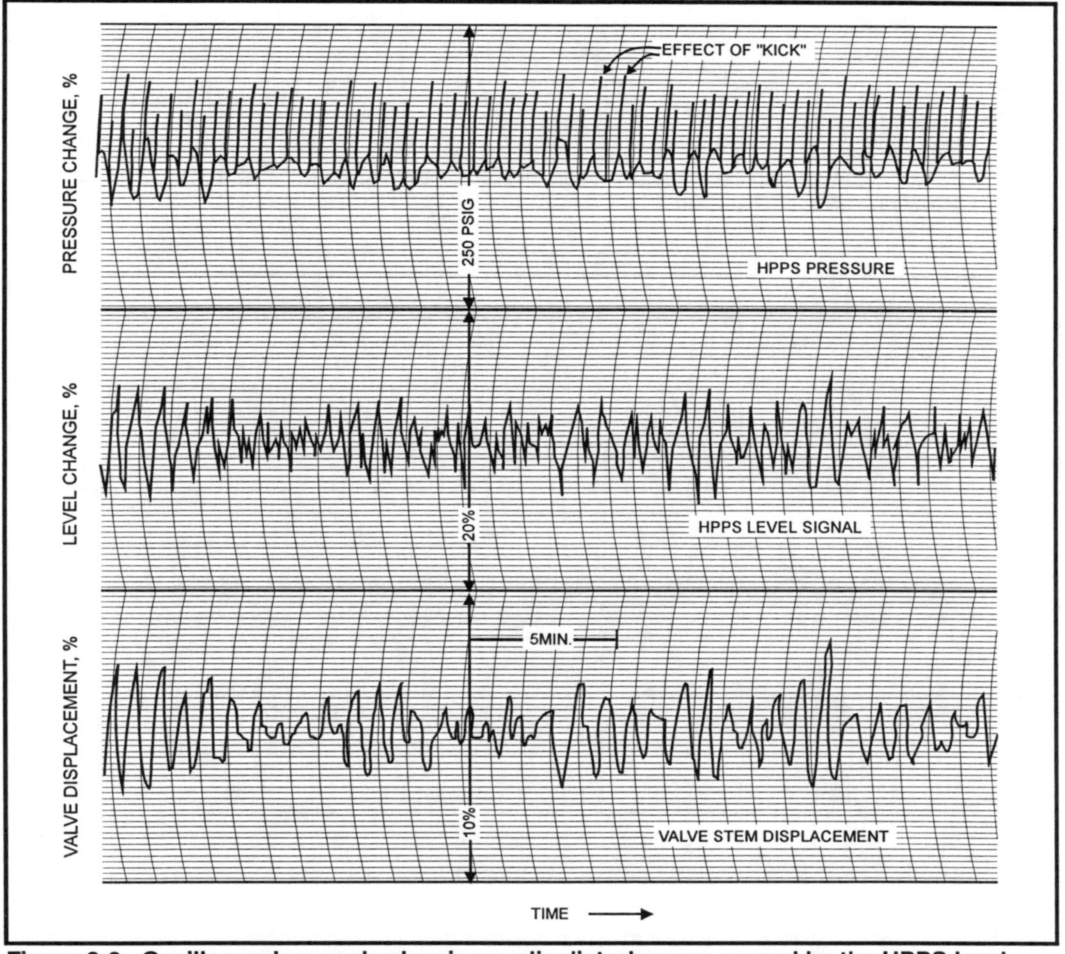

Figure 3-2. Oscillograph records showing cyclic disturbances caused by the HPPS level control system

These records suggested that level controller gain was probably excessive, causing excessive valve stem motion. Also, at peaks of valve displacement, light phase from the HPPS probably escaped causing rapid diminutions in HPPS pressure. The changes in pressure at various points in the system caused by oscillations of HPPS level control system are shown in Table 3-1.

Upon placing the HPPS level controller in the manual mode oscillations disappeared. A typical oscillograph record appears in Figure 3-3, demonstrating that oscillatory behavior was caused by the HPPS level control system. Performance of the level system in both the automatic

Table 3-1. Magnitude of Pressure Changes Caused by Oscillatory Behavior of HPPS Level Control System

Pressure, PSIG	HPPS	Upstream of LP Recycle Valve	Upstream of LP Recycle Meter	Booster Suction	Make-up B Discharge	Hyper A Suction
Maximum	3575	22.3	11.7	10.6	3420	3365
Minimum	3475	11.8	7.8	6.8	3220	3290
Difference*	100	10.5	3.9	3.8	200	75

*Double amplitude
Both low-pressure recycle and booster suction control valves 42% open.
Make-up valve 27.9% open

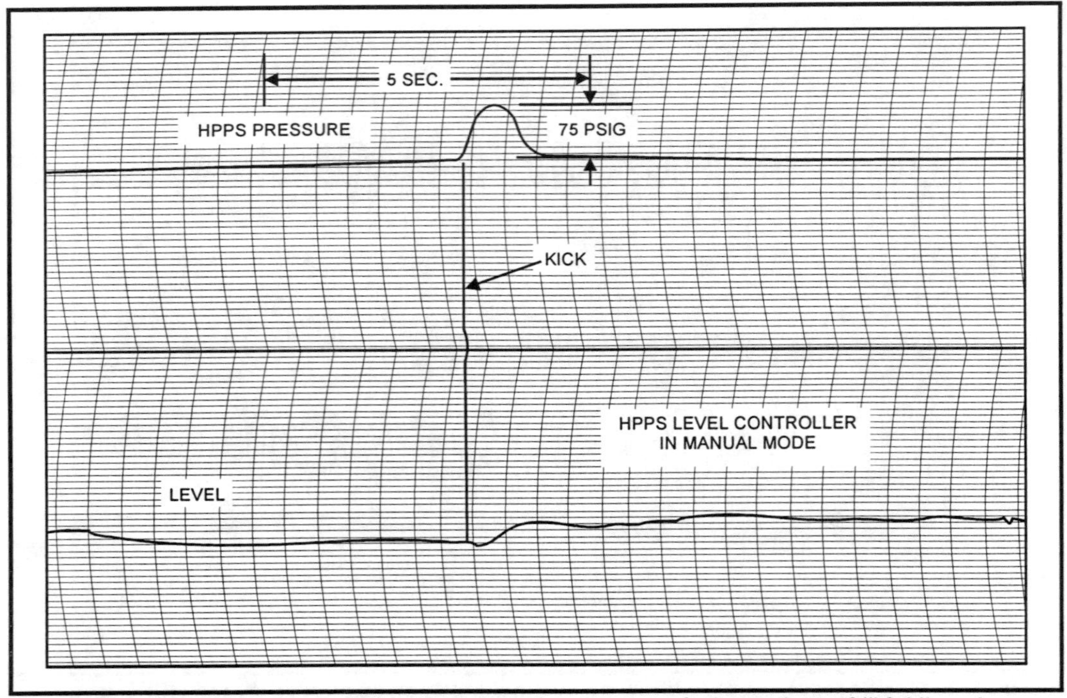

Figure 3-3. Oscillographs verifying that disturbances were independent of "kick" cycle

and manual mode is shown in Figure 3-4, demonstrating unequivocally that this control loop was the primary cause of undesirable oscillatory behavior. Absence of oscillations in HPPS pressure (upper right-hand record) substantiates the assumption that light material, along with liquid, was escaping from the HPPS when the control system was active.

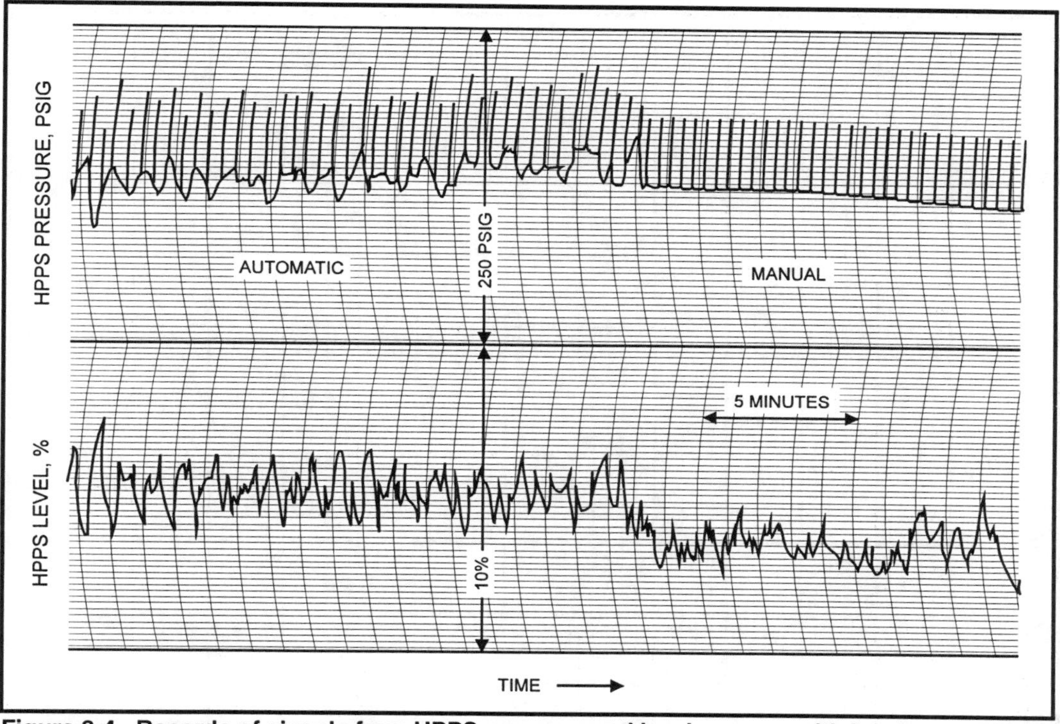

Figure 3-4. Records of signals from HPPS pressure and level sensors with level controller in manual mode

Some steady-state observations were next obtained while the valve was cautiously moved from one position to another, with the HPPS level control in the manual mode. Results are shown in Figure 3-5. Full valve stroking was not feasible with the system in operation, therefore establishing fully closed and fully open positions was not possible. The abscissa in Figure 3-5 refers to deviations in the record of valve stem displacement from a reference position denoted as "zero." Deviations are divisions of pen deflection on the oscillograph chart; decreasing deviations correspond to valve closing. Moreover, the flow rate of low-pressure recycle is indicated merely as the pressure drop, in inches WC, across the orifice plate of this flow sensor. Since fluid properties were uncertain, estimates of actual mass flow rate would have been most difficult to ascertain.

These steady-state (or static) data (shown in Figure 3-5) demonstrated dependence of important process variables upon the flow rate through the HPPS level control valve. Note especially that LPPS pressure appeared to increase almost exponentially as the level control valve was progressively opened.

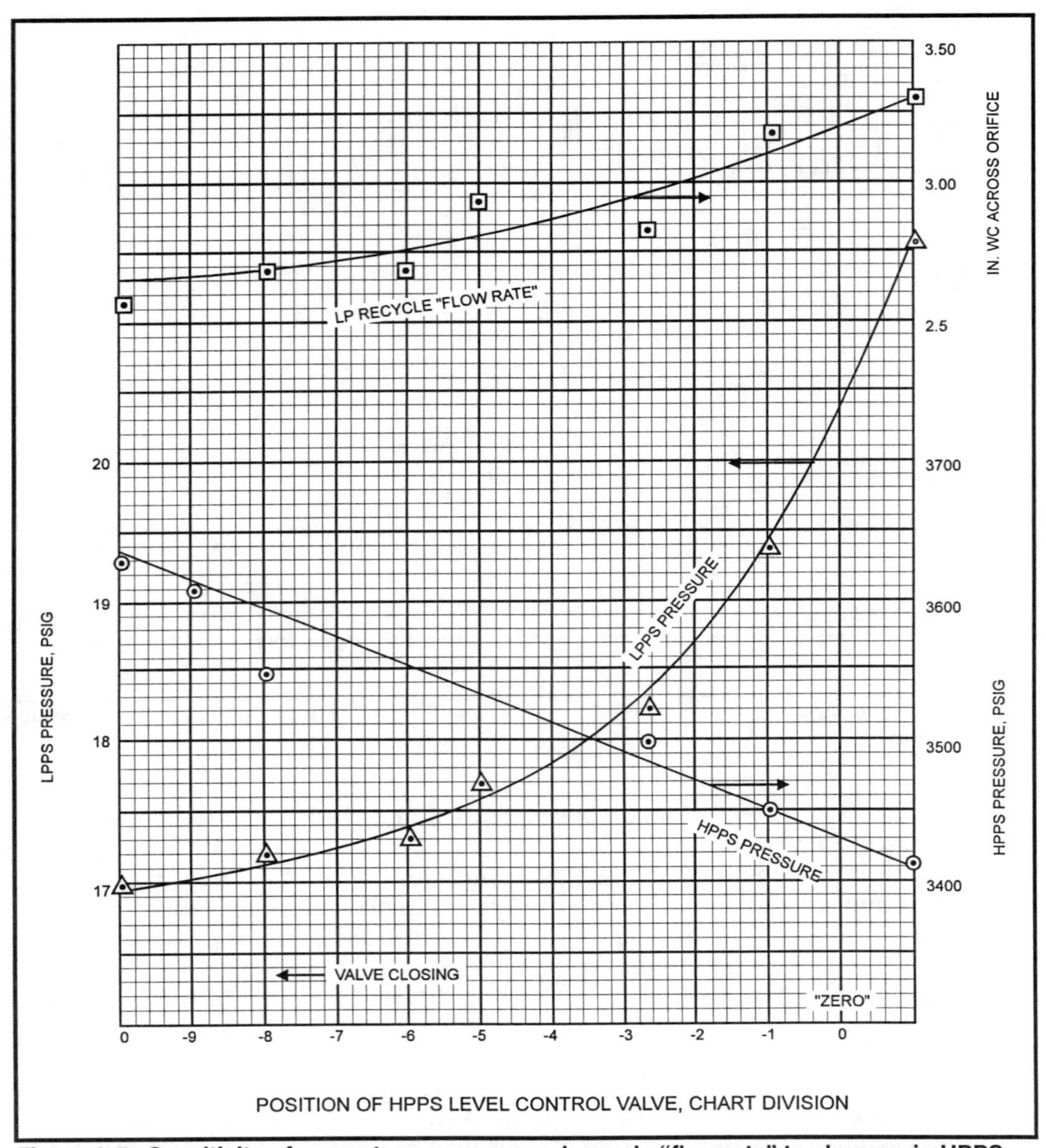

Figure 3-5. Sensitivity of separator pressures and recycle "flow rate" to changes in HPPS level control valve displacement

Some Dynamic Tests

Before discussing further details of the HPPS level control system, the results of several dynamic tests are presented. These tests were conducted to determine the transient response of pertinent variables to inserted disturbances. In all these tests the input variable, or disturbance, was changed in a pulse-like manner and responses recorded as oscillographs. A description of how such tests should be conducted and methods of data reduction are fully described and illustrated in Appendix B.

To execute a pulse test an independent variable is forced in a pulse-like manner and responses of dependent variables recorded. A pulse, as implied here, is a function of time that departs from some steady-state value for a finite period, called pulse duration, before returning to the initial value. Dependent variables usually behave similarly, although responses may not return to initial values. Indeed, neither input nor outputs need to close in order to retrieve the desired information, but special precautions and data processing procedures are required in these situations. (See Appendix B).

Three subsystems were tested in this manner with the following results.

1. Response of the booster compressor suction pressure to changes in displacement of the booster suction control valve.

   Time histories are shown in Figure 3-6, along with the frequency response derived therefrom. The dynamic relationship is essentially first-order, *i.e.*,

$$\frac{\text{booster suction pressure, psi}}{\text{displacement of bypass valve, \%}} = \frac{2.15}{1 + 14.3s}$$

   A delay, or dead time, of about 1 second was observed.

2. Dynamic relation between displacement of low-pressure recycle control valve and LPPS pressure.

   Time histories and frequency response are shown in Figure 3-7. Again, a first-order relationship fits the data quite well, this being

$$\frac{\text{LPPS Pressure, psi}}{\text{displacement of LP valve, \%}} = \frac{4.32}{1 + 33.4s}$$

   The absence of pure delay time should be noted.

   These results suggested that both pressure controllers used for the indicated service need impart only proportional action. Closing the loop with unity feedback will, in both instances, produce another first-order system, the response of which can be made as rapid as needed by merely increasing the controller gain. Insistence upon introducing an integration in the controller function, in such cases, produces a second-order closed-loop with subsequent oscillations if improper controller adjustments are used.

For readers who maybe interested in the above conclusions we offer the following analysis.

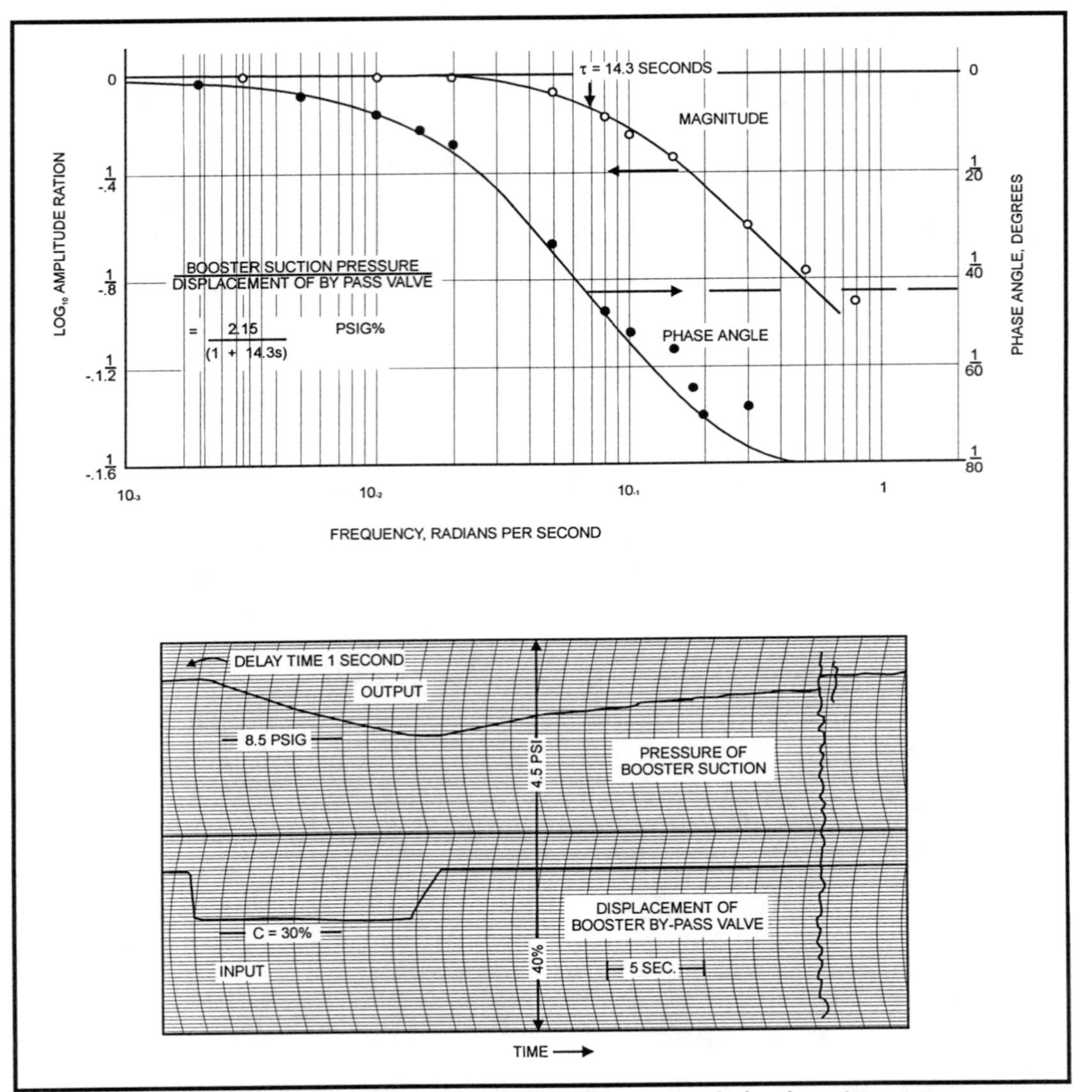

Figure 3-6. Time histories and derived frequency response relating booster compressor suction pressure and bypass valve displacement

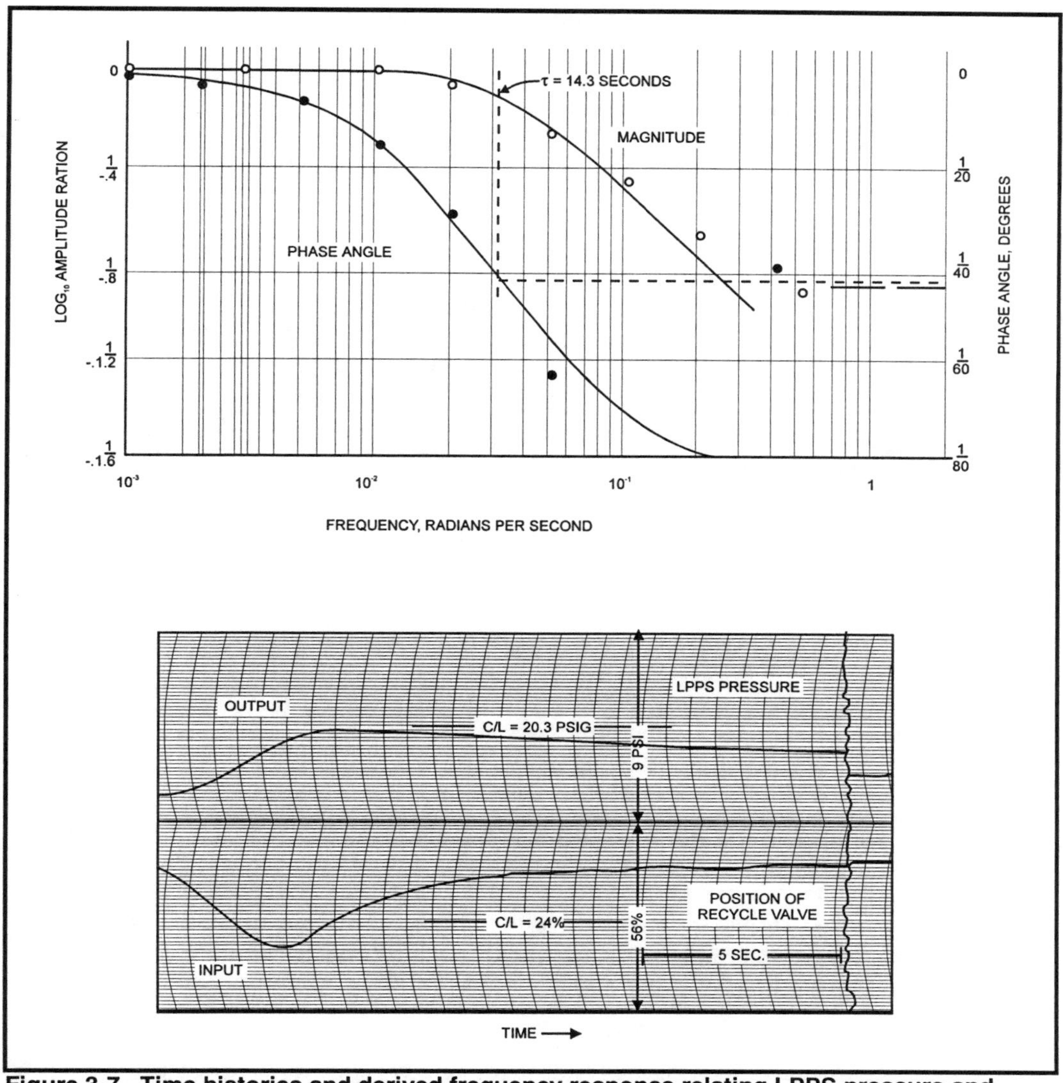

Figure 3-7. Time histories and derived frequency response relating LPPS pressure and displacement of LP recycle valve

The typical block diagram of the closed-loop system appears below, where R is the setpoint or reference and C is the output, or desired value.

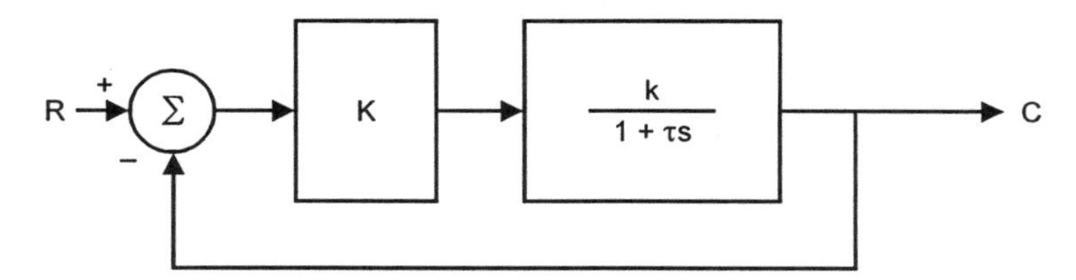

From the above the relations below follow:

$$C/R = \frac{\frac{Kk}{1+\tau s}}{1+\frac{Kk}{1+\tau s}} = \frac{Kk}{1+Kk+\tau s} = \frac{1}{1+\frac{\tau}{1+Kk}s} \rightarrow 1$$

as $K \rightarrow \infty$.

As the controller gain is increased, it is seen that the time constant of the closed-loop decreases and the closed-loop gain approaches unity. The offset, or steady-state error, will be determined by the magnitude of the controller gain, which, in theory, can be made arbitrarily large. Of course, if dead time is present or additional dynamics appear in the system, a purely proportional control may not be satisfactory.

3. Relation between reactor II inlet pressure and the displacement of the "kick" valve.

   The pulse test data and frequency response are shown in Figure 3-8. The result is

$$\frac{\text{reactor II inlet pressure}}{\text{displacement of kick valve}} = \frac{5.08}{1+2.63\,s}, \text{psi}/\%$$

   This relationship has no bearing on control but is presented in order to suggest the possibility of a relationship between the apparent time constant, gain, or shape of the frequency response and quality of final product. (The dynamic relationship between reactor pressure gradient and "kick" valve stem position might prove to be a more sensitive indication of product quality.)

   This speculation assumes that changes within, and the pressure gradient across, the reactor reflect changes in the properties of the material within the reactor. Thus, for example, the viscosity of the material within the reactor will influence the pressure gradient and, also perhaps, the pressure at the inlet to the "kick" valve. The Fourier transform of either signal is one way to obtain a description of the frequency content of the time history. The suggestion is made that perhaps the amplitudes and phase angles at various frequencies, derived from the transform, might correlate with the properties of the final product. If such relations were found, perhaps conditions at the reactor could be adjusted more promptly than awaiting the testing of the finished product, thus reducing the amount of off-specification material.

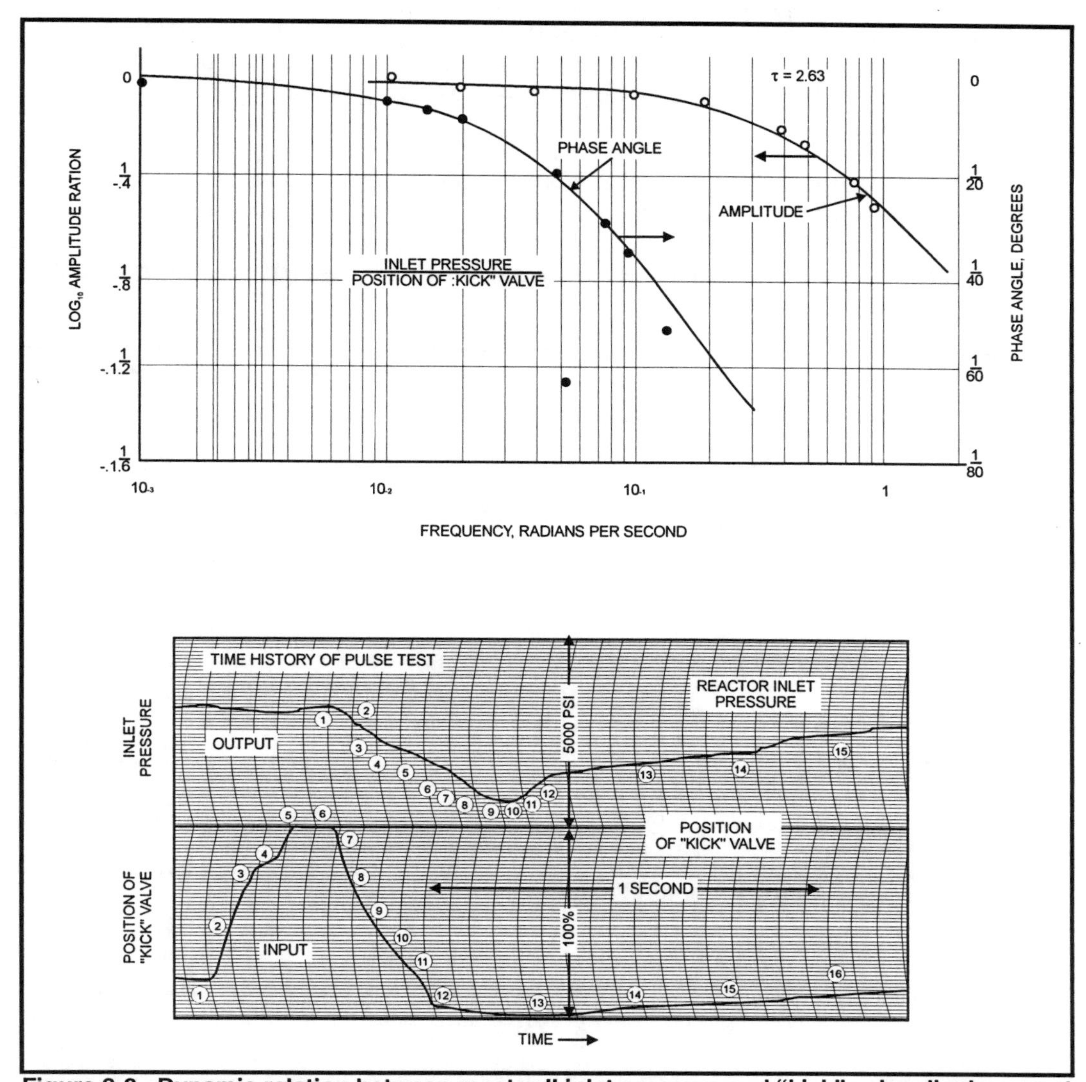

Figure 3-8. Dynamic relation between reactor II inlet pressure and "kick" valve displacement

Analysis of the HPPS Level Controller

From data previously presented, the HPPS level controller was suspected as being the source of the undesirable oscillations. For this reason a study of the circuit diagram of this component became necessary. One may ask why this analysis was not immediately undertaken. There were three reasons for delay: first, avoiding a common tendency to immediately suspect the most complicated source of trouble; second, the information described above was conveniently obtained at this juncture in the study; and third, the controller circuitry was quite complicated and translating ponderous manuals written in a foreign language was a forbidding task.

A small, but pertinent, portion of the very complicated circuitry is shown in Figure 3-9. This is the circuitry by which the gain of the controller is adjusted, which parameter appeared to be excessive in the HPPS level controller.

The sensitivity, or gain, of the manual adjustment was reduced by replacing the 2K potentiometer (shown as R_F) by a 1K, 10-turn potentiometer. (To preserve the total resistance in the path between two voltage sources a 1K resistor was inserted.) Although the HPPS level valve

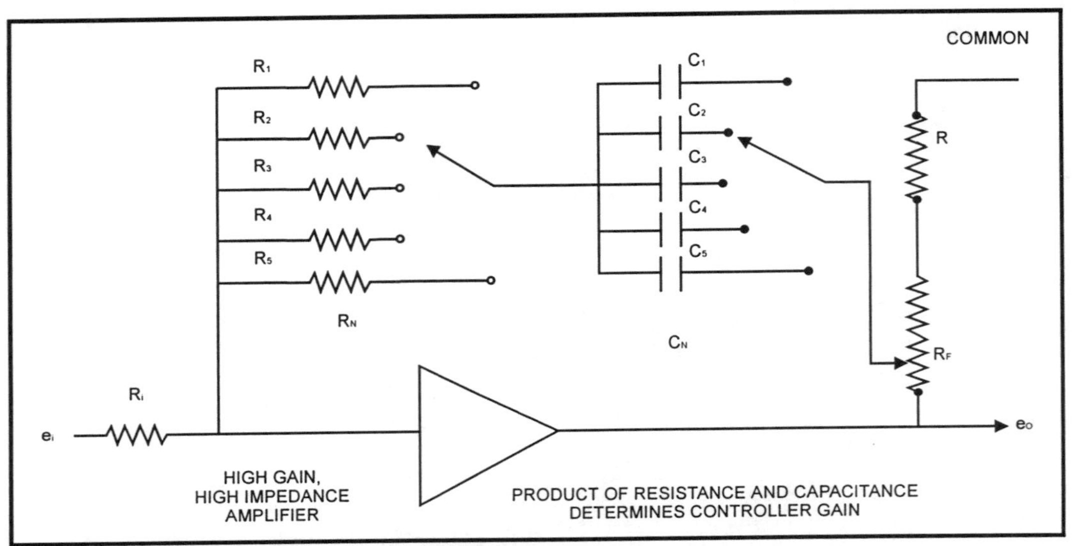

Figure 3-9. Circuit for adjusting controller gain

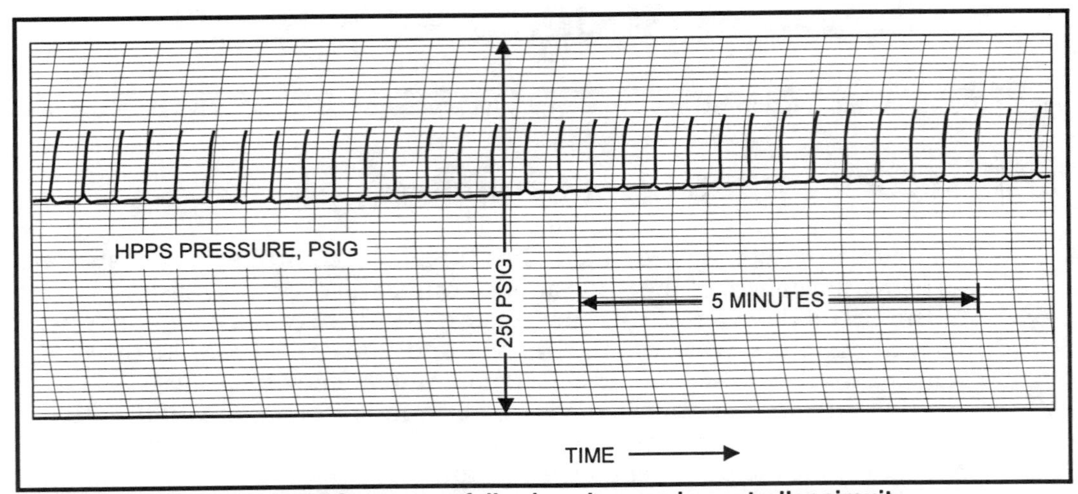

Figure 3-10. Record of HPPS pressure following changes in controller circuit

could not be opened more than 50%, this portion of its stroke could be allocated to about 8 turns of the potentiometer. With these changes the HPPS level control valve could be easily positioned within this range and, as shown in Figure 3-10, when this controller was again placed in service, the oscillatory conditions were eliminated.

The controller circuit (shown in Figure 3-9) produces the function:

$$e_O(s) = \frac{1 + (R_N + R_f)C_f s}{R_i C_f s} e_i(s)$$

Validity of this relationship was tested by experimental measurements using a step change in the voltage applied to the controller input. The output was the expected step change, followed by a ramp, the slope of which is indicative of the rate of integration. Table 3-2 presents the results of the calibration of this controller.

Table 3-2. Dynamic Characteristics of the Level Controller

Capacitor No.	Capacitance Mfd	Time Constant Seconds
1	4.7	28
2	10	55
3	20	105

These results applied to all controllers of this kind used to position hydraulically actuated valves installed on this process.

Other electronic controllers providing proportional plus integral action were also calibrated in similar fashion with results shown in Table 3-3.

Table 3-3. Calibration of Reset or Integral Action of Controllers

Reset Dial Reading, Repeats per Minute	Experimentally Determined Integration Time Constant, Minutes
0.75	3.80
0.30	0.83
0.10	0.40

We cannot account for these results differing from those associated with electronic controllers manufactured in this country, where "repeats per minute" is approximately equal to the reciprocal of the integration, or reset, time. This information is presented merely to emphasize the need to know the validity of the adjustable controller parameters, if a control system is to be designed.

Comments on Valve Performance

In other studies, performance of valves used for control purposes were discovered to conform to a relation of the form:

$$W = Ke^{kz}\sqrt{\Delta P_v \rho_v}$$

where:

W	=	mass rate of flow through the valve
e	=	natural logarithm base
ΔP_v	=	pressure drop across the valve
z	=	position of the valve,% open
ρ_v	=	density of flowing fluid, and

K and k are empirical constants, where k depends upon specific design of the inner valve (*e.g.*, plug, butterfly, double seated, etc.), and K depends on nominal size.

In this study a Masonelian Camflex® valve was used to regulate the overhead material, largely ethylene, flowing from the LPPS. Since temperature and pressure were measured, the flow rate through the flow meter could be computed with some confidence. A plot of $W/(\sqrt{(\Delta P_v \rho_v)})$ on a log scale vs. valve displacement, z, in percent, gave the results shown in Figure 3-11. These results show that convenient relationships can be found to describe valve performance and suggest that, in some instances, valves can be used as sensors of flow rate.

Suggested Control Strategies for this Polyethylene Process

A significant requirement for control of this process is regulation of gaseous inventory. Material leaves the process both as liquid product and as gaseous purge. To maintain gaseous inventory and supply fresh reactant, ethylene is added at a pressure somewhat above the pressure at the point of addition.

The deficit of light material could best be detected at the front end of the process, for example, near the "booster" compressor suction. However, rejection of purge material from this area poses problems unless the flow rate of purge is known with considerable precision.

The original control scheme utilized make-up compressor discharge pressure as an indication of the need for ethylene addition. Thus, as this pressure decreased, more ethylene would be added to the suction of these compressors. Some concern existed that fresh ethylene might be included in the purge gas with this arrangement, hence a valve (see valve A in Figure 3-12) was installed to provide an appropriate pressure gradient. This bypass valve was positioned manually.

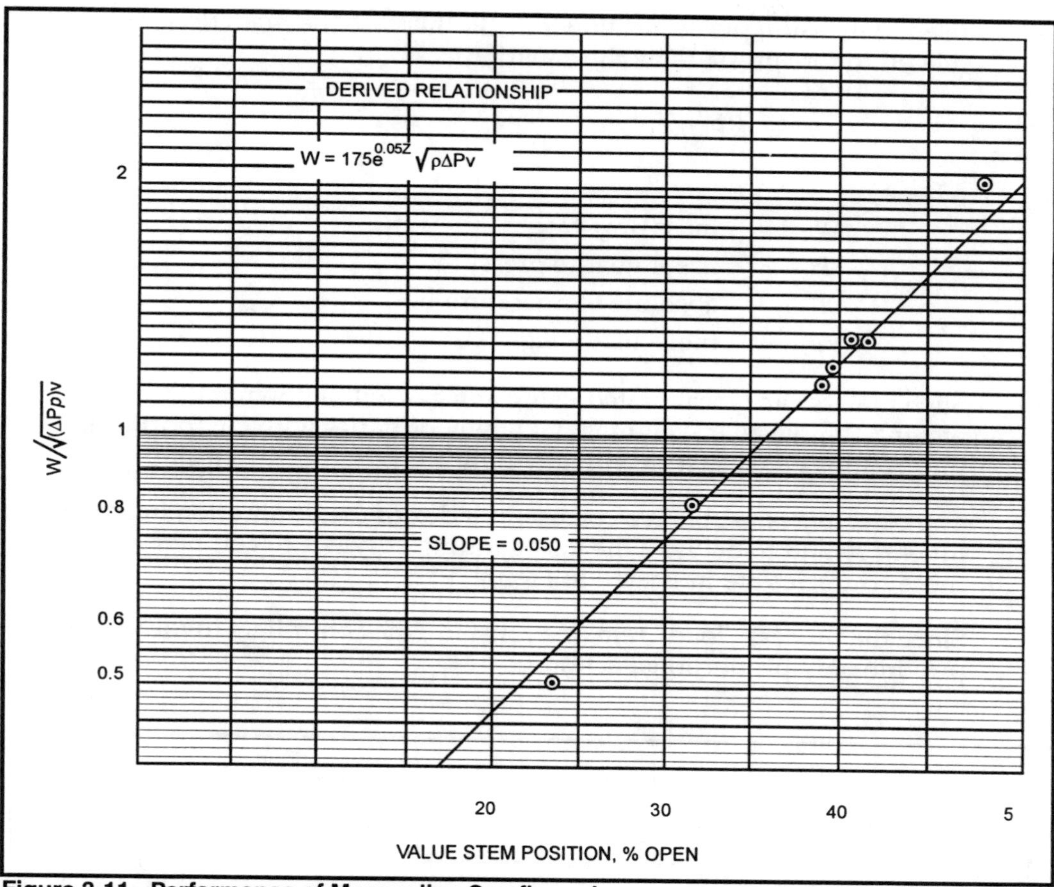

Figure 3-11. Performance of Masonelian Camflex valve

Two alternative strategies are suggested.

1. Measure the flow rate of gas from the LPPS and increase ethylene addition in proportion to the diminution in the flow rate of this gas.
2. Measure the flow rate of purge gas and add ethylene in proportion to this rate.

In both cases accurate indices of flow rate must be available. These can be pressure gradients across properly installed orifices (or preferably flow nozzles or Venturi tubes) although simpler techniques might suffice, for example, merely measuring valve stem positions.

Provided the upstream, or forepressure, is about 2-1/2 times the absolute downstream pressure, an orifice (or nozzle) operating critically might be used to measure purge gas flow rate. In this case the mass flow rate is related to fore pressure by the equation

$$W = KP\sqrt{MW/T}$$

where

W = mass flow rate
P = absolute fore pressure
T = absolute temperature
MW = molecular weight

Such meters can be calibrated using any gas with known molecular weight. If molecular weight is unknown, or varies with composition, uncertainties arise just as with any flow meter.

The objective is to obtain a consistent index of the diminution in gaseous inventory and to devise a means of adding ethylene to compensate. Both these strategies are indicated in Figure 3-12.

Since the process "load" remains constant, the controllers involved need provide only proportional action.

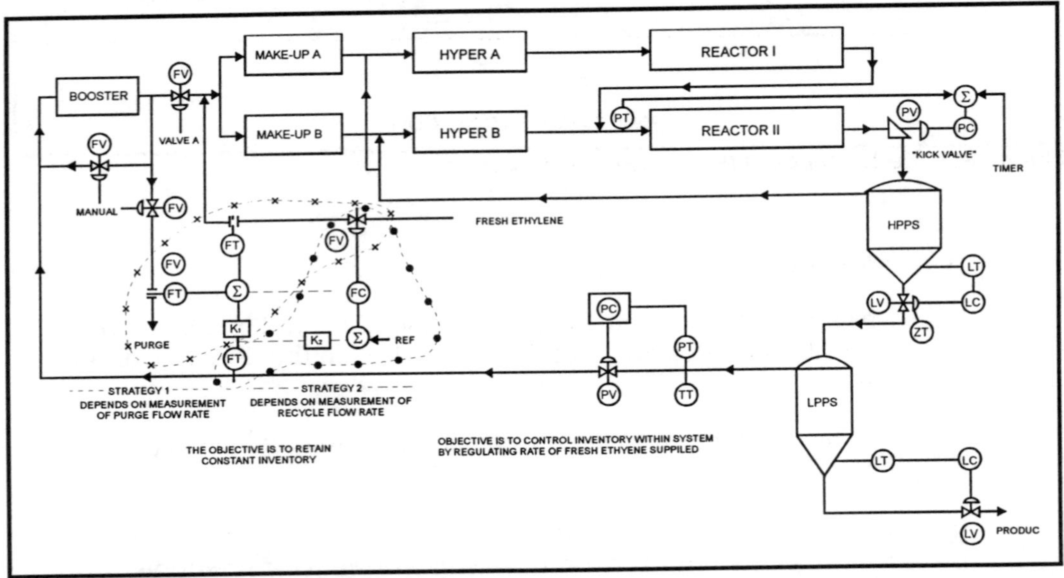

Figure 3-12. Proposed control strategies

What was Learned from This Study

- Seemingly difficult problems sometimes have remarkably simple solutions. In this case study the replacement of a resistor in the circuit of an electronic controller was the major requirement.
- The dynamic response characteristics of several process subsystems were found, and in each case the relationships were simple: first-order, with some pure delay in one instance.
- The dynamics of these subsystems clearly indicate that they may be regulated by controllers providing only proportional action.
- The performance of control valves can frequently be described by simple, yet very useful, relationships.
- For this process, preservation of inventory is not only necessary, but appears to be easy to monitor and control. Two strategies are suggested by which this might be accomplished.
- However simple the ultimate solution to a processing problem may be, a considerable investment in time and effort is usually required for its discovery.

4
Control of pH

Problems initially perceived as difficult occasionally turn out to have simple, often obvious, solutions. One newly introduced to a problem is often persuaded to agree with those who have previously struggled with it, and who are presumed to be better informed. But this is only to excuse one's lack of curiosity. The case presented here illustrates this behavior, and what appeared to be an intractable situation proved to be relatively easy to rectify with very little in the way of process changes or expensive revisions.

This study provides a valuable lesson and suggests a reasonable protocol to follow in any process study. When faced with a processing problem, first scrutinize every detail of all processing apparatus, taking nothing for granted. Do this before making any commitments or launching into an extensive experimental study. It is in the details that troubles lurk.

In preparing for this investigation, the above tenets were not strictly heeded. As a result, the major source of difficulty was not identified until considerable effort had been expended. This preliminary work, while having no direct bearing on the solution to the problem, did, however, produce useful information and served to direct attention to the principal cause of poor pH control.

The Process

The purpose of the process was to furnish a subsequent operation with a continuous supply of an aqueous solution at constant pH. This was to be accomplished by the addition of dilute hydrochloric acid (HCl) at a point somewhat removed from the point of pH measurement. The neutral solution, prepared elsewhere, entered at a pressure sufficient to sustain level in reservoir A (see Figure 4-1). From reservoir A the solution is pumped through the control valve and heat exchanger into the cone bottom of vessel B. Minor additions of a more highly concentrated solution may be added to this vessel; such additions, however, are of little importance in this study.

The solution in vessel B overflows into the annular weir encircling the inside circumference of vessel B and discharges through a single outlet from the trough at the point indicated. Attention is directed to the details of the path of the overflow as it descends from the weir to vessel C some 20 feet below. The overflow, collecting in the annular trough about 8 inches deep, enters a short circular conduit about 6 inches in diameter. It then progresses horizontally for about 10 inches before descending vertically for about 20 feet. After passing through a magnetic flow meter, it is directed horizontally through a tube-and-shell heat exchanger in which the solution may be heated. Emerging from the heat exchanger, the material flows vertically, perhaps 8 feet, before entering the top of vessel C through a short horizontal section.

It is to this latter vessel, having a capacity of about 1000 gallons, that HCl (about 30%) is added with the intention of controlling the pH of the solution near the point of its discharge into vessel D. The level in vessel D was controlled by regulating the flow rate through the control valve located in the line supplying solution to vessel B. The controller was pneumatic, providing proportional plus integral control functions. With two intervening vessels between valve and level sensor, this arrangement was immediately suspected of contributing excessive signal delay, as well as possible deleterious dynamics. (The effluent from vessel D was carefully metered to a number of items not shown nor of concern here.)

The Problem

Great difficulty had been experienced in achieving control of pH as measured by a pH electrode positioned near the surface of the liquid in vessel D. The pneumatic output from this sensor was the input to a pH controller, which, in turn, controlled the addition of aqueous HCl.

No semblance of control had been achieved, acid addition was virtually on-off, and the pH usually varied from one extreme to the other in a very erratic manner. Furthermore no meters in the immediate vicinity of the processing facilities shown in Figure 4-1 were available for measuring the flow rate of the fresh solution. Since difficulties were experienced in introducing the HCl solution, the magnetic flow meter installed in this line was occasionally bypassed in order to obtain an adequate flow rate of acid. The magnetic flow meter in the vertical line leading from vessel B was not in service because of suspected malfunction.

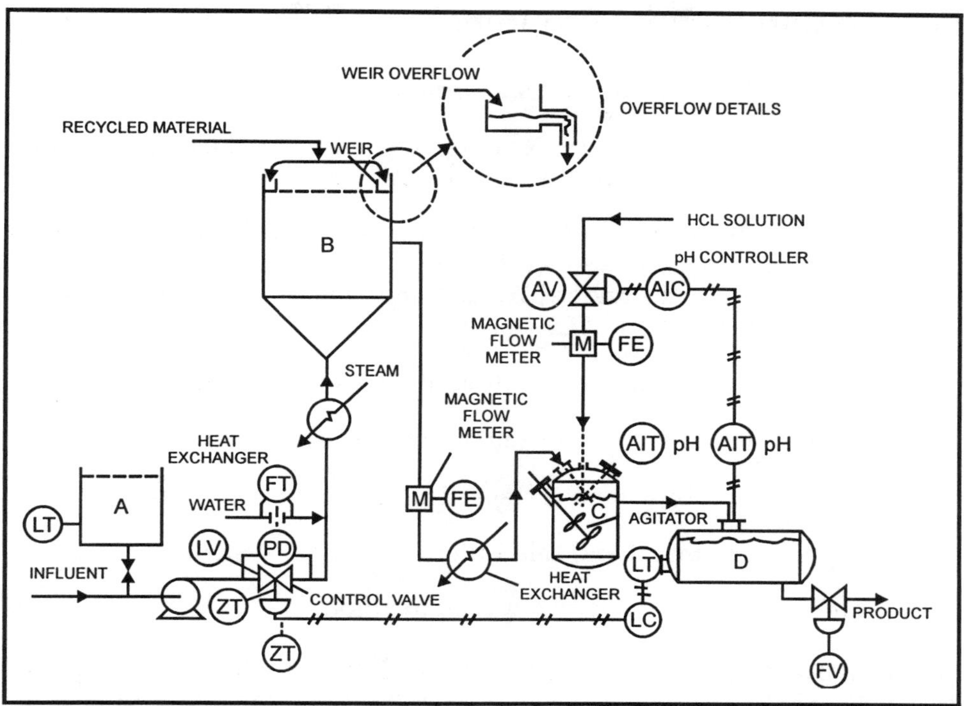

Figure 4-1. Process flow diagram with instrumentation and points of measurement

Plant Changes Prior to Testing

As intimated previously, the test crew overlooked a few items concerning the nature of the problem and, consequently, anticipated the need for more information than was ultimately required. Hence, while awaiting the arrival of special test apparatus considered necessary, some work (listed below) was requested from plant personnel.

1. The magnetic flow meters were serviced and calibrated using their internal signal-generating capability. This included the meter in the overflow line from vessel B.

2. In addition to the more readily available signals, special provisions were made for measuring the following items:

   a. Pressure drop across the control valve used in conjunction with the control of level in vessel D.

   b. Position of the stem of the above valve.

   c. Pressures of air to valve positioner and on the diaphragm of this valve.

   d. Pressure at the bottom of vessel A, a measure of level therein.

   e. Level in vessel D.

   f. The pH of the overflow from vessel C.

Preparation for Test Program

Among the preparatory tasks completed by the test crew were:

- Experimental studies of the response of pH electrodes and associated signal measuring systems.
- Calibration of all special pressure and differential pressure strain-gage-type transducers, using either a dead weight prover (above 25 psig) or a standard Wallace and Tiernan calibrated pressure gage (up to 200 in. WC).
- Assembly and calibration of signal filter circuits.
- Preparation and installation of signal cables.
- Assembly of valve stem displacement transducers and circuits.
- Adjustment and calibration of the data system.

The first of these deserves special mention.

Response Characteristics of pH Electrodes and Measuring Systems

Transient response data were obtained from various pH-sensing configurations in order to gain some concept of the dynamical problems these components might impose, insofar as pH control was concerned. A special high-impedance module accompanying the oscillograph was used to measure the output of the pH sensors. Figure 4-2 shows the method of measuring the output from a typical component used in conjunction with a pH electrode. The resistor R may be 4 or 5 ohms. This millivolt signal to the oscillograph amplifier was preferred to the pneumatic output from the input transducer. (Selected results are shown in Figure 4-3.)

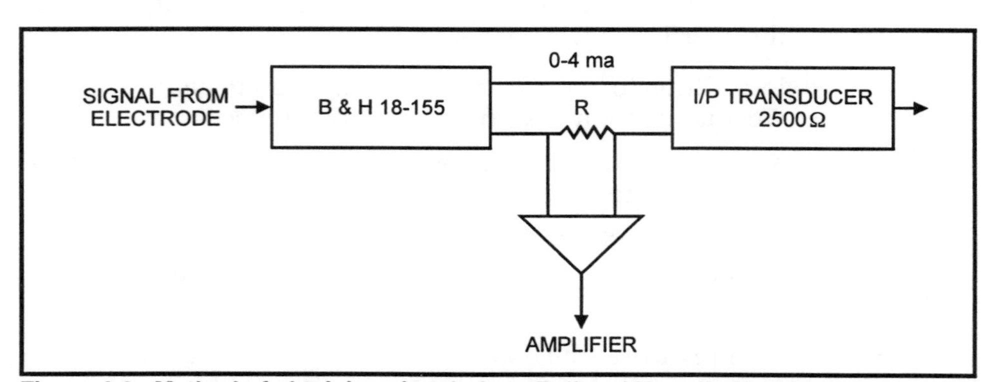

Figure 4-2. Method of obtaining signals from Bell and Howell pH converter

The upper left-hand time history shown in Figure 4-3 is the response of the electrode, which occurs when the influent changes from water (pH 6.45) to a buffered solution having a pH of 10.7. The flow rate was 0.68 L/min. through the Beckman flow cell. This response is strongly first-order, the apparent time constant being about 4.5 seconds.

Upon reversing the direction of change (right-hand record), the response is not only much slower but is distinctly higher order, more like second-order. The 63.2% point is attained in about 46 seconds, 10-fold greater than the previous response.

The next lower set of records was obtained when the pH was changed from 9.2 to 10.7 and vice-versa. While the time constant for the increasing pH (left-hand record) is greater then the previous record, changing from a pH of 10.7 to 9.2 (right-hand record) shows a reduction in the approximate first-order time constant from 46 to 29 seconds.

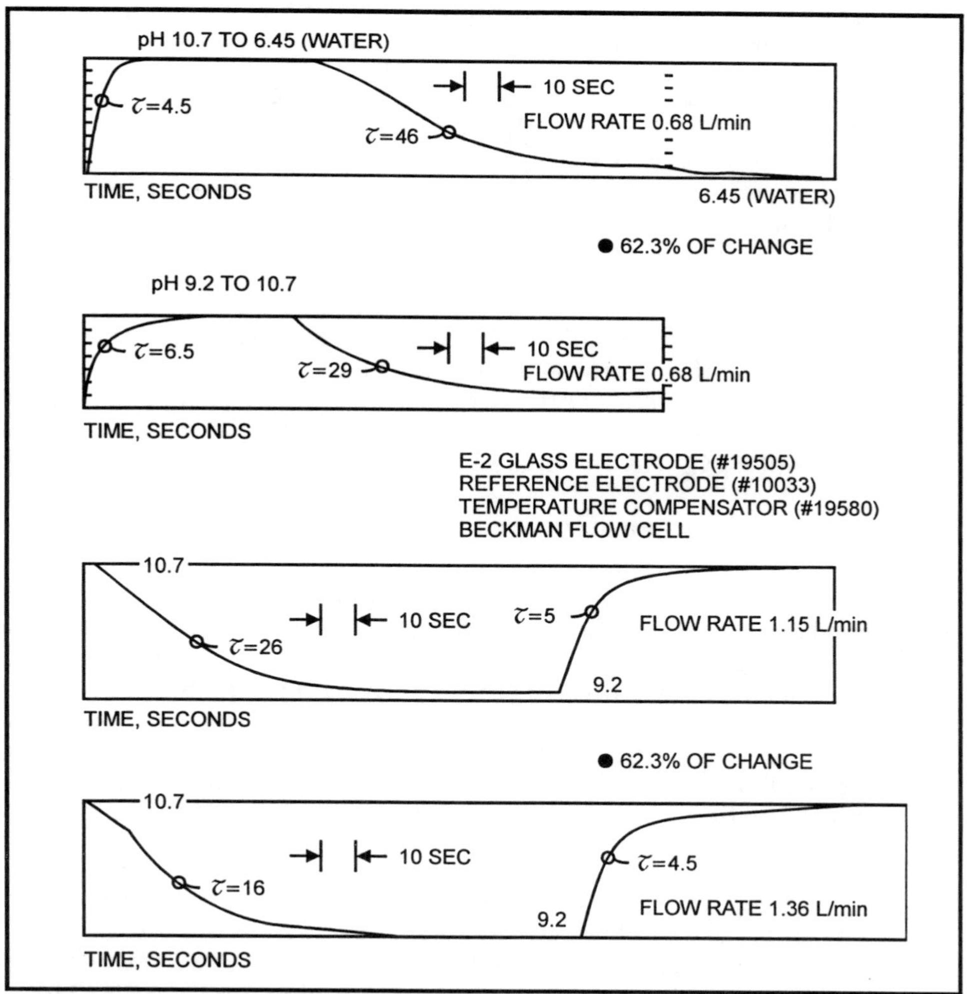

Figure 4-3. Transient responses of pH sensors

The lower records in Figure 4-3 are included to demonstrate the dependency of response on flow rate for the same pH changes shown previously. As before, the response is slower for a decreasing step change, and there is some suggestion of higher-order dynamics than first. Table 4-1 presents data at three flow rates, indicating that a flow rate of at least 4 liters per minute through the cell should be used.

Table 4-1. Effect of Direction of pH on First-Order Response of pH Sensor

	pH Change	
Flow rate, L/min.	**9.2 to 10.7**	**10.7 to 9.2**
0.68	6.5 sec.	29 sec.
1.15	5.0	26
1.36	4.5	16

The great dependency of response upon the direction of pH change is the most important result of this work.

This writer has no ready explanation for the observed performance of these pH sensors. One might speculate that response is determined by the ion species, their concentration in the vicinity of the electrode, and on the relative tenacity of the ions to adhere to the electrode surface. Differences in these properties could account for the responses being dependent both upon the direction of change and the velocity of the liquid flowing past the electrode.

A titration curve was also obtained on a sample of the liquor being processed prior to any acid addition. Figure 4-4 shows the region in which the sensitivity of the pH sensor was likely to change drastically, in this case between pH 5 and 2.5. This behavior, conducted at a temperature of about 25°C, was anticipated to vary somewhat with temperature.

Preliminary Observations

The usual testing policy is to select as many variables for measurement as are considered significant and to obtain a set of records documenting performance. The aim is to evaluate existing control systems and gain some concept of processing difficulties and malfunctions. In this case, this procedure verified extreme uncontrolled variations in pH and in certain flow rates, in particular large variations in flow rate of overflow from vessel B into vessel C, as indicated by the output of the reactivated magnetic flow meter in this service. (Selected records are shown in Figure 4-5.)

Obviously the entire system was never at steady state, with no indications of control of any recorded variable. Thus, there was difficulty in identifying the source of the oscillations. In view of what was to be learned later, it is safe to conclude that a major driving input was the change in flow rate of overflow from vessel B into vessel C. The level change induced in vessel D (as recorded in the third record from the top of Figure 4-5) is responsible for causing the position of the level control valve to oscillate from fully closed to fully open (as seen in the top record of Figure 4-5.)

Upon noting the large and erratic variations in the flow rate of the material entering vessel C (as shown in the lower record in Figure 4-5), the test crew was, at last, motivated to view the inside of this vessel through a somewhat inconveniently located access port. Immediately the reason for the oscillations in the flow rate from vessel B became clear. The liquid descending from vessel B to vessel C was subject to variation in density due to entrained air! Thus, as air became progressively entrained, the fluid head was finally reduced to the point at which flow virtually ceased. As air was released and density increased, flow resumed until entrained air again reduced the density, and the cycle was repeated. These large variations in flow rate could be recognized by simply viewing the discharge from vessel B into vessel C.

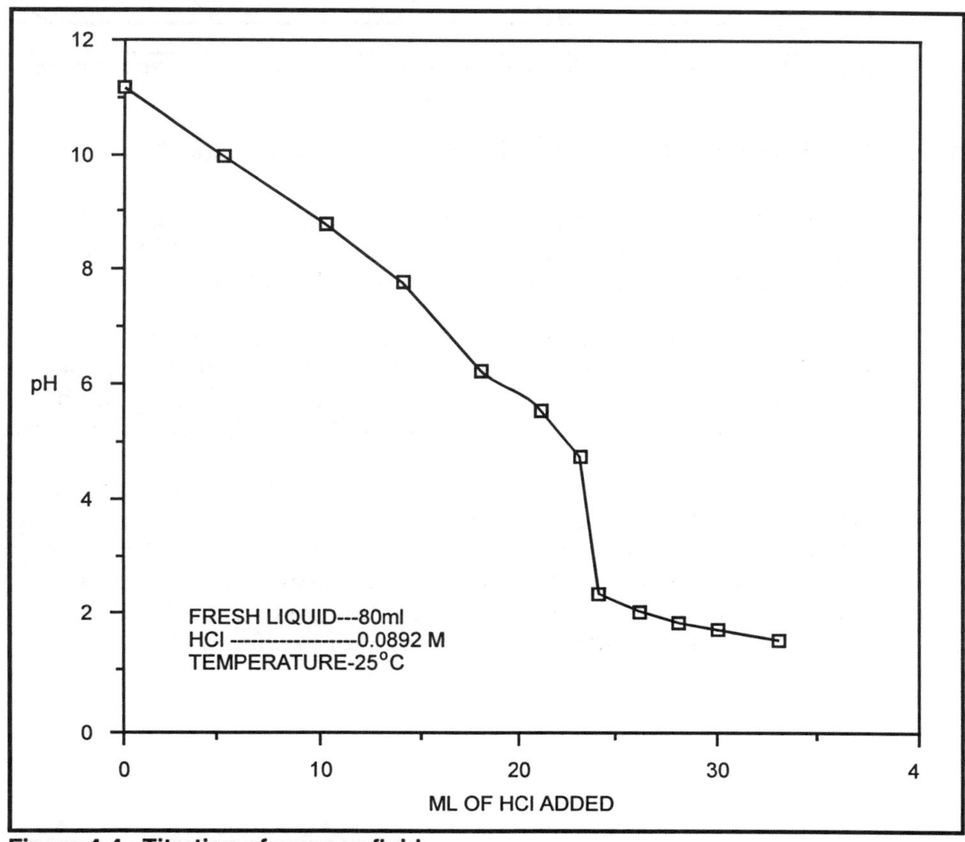

Figure 4-4. Titration of process fluid

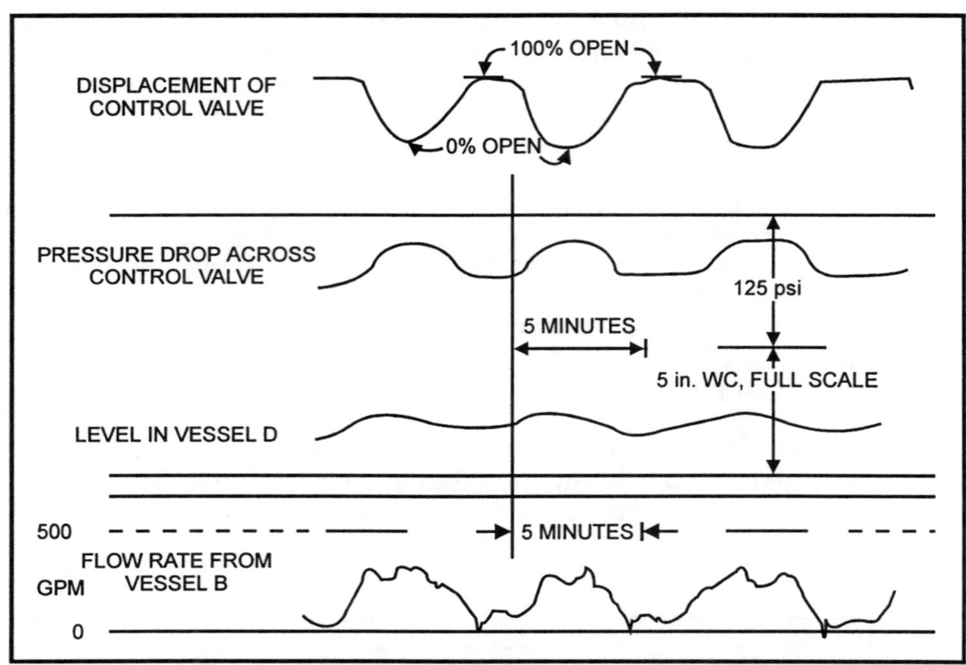

Figure 4-5. Selected oscillograph records showing unsteady performance

Effect of Changing Level Controller Parameters

The records shown in Figure 4-5 were typical of existing conditions. The control valve, actuated by the level controller associated with vessel D, oscillated between being fully open and fully closed.

It was difficult to separate cause and effect because of the inherent delay time between the different signals. While it was logical to suspect that the (large) changes in flow rate from vessel B were solely responsible (see the lower time history in Figure 4-5), the level controller was also suspect.

At the time this record was obtained, the "proportional band" and "reset" adjustments of the level controller were 10% and 0.3 minutes, respectively. Upon increasing the reset adjustment to between 5 and 10, the records shown in Figure 4-6 were obtained. The control valve, while still cycling, does not fully close. Clearly the controller was a contributor to the cyclic behavior. An explanation for this behavior is offered later.

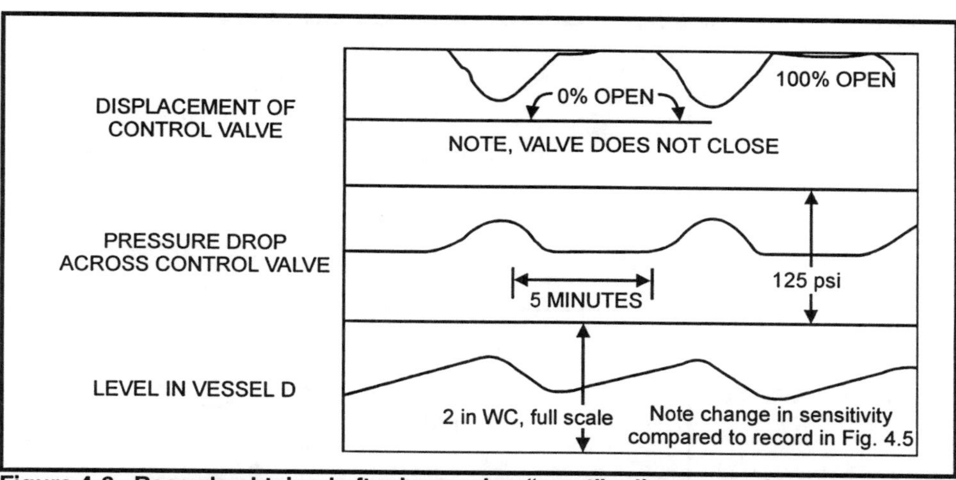

Figure 4-6. Records obtained after increasing "reset" adjustment of level controller

Improvising a Flow Meter

Because there was no meter for measuring the flow rate of material flowing from vessel A to vessel B, the control valve in this line was calibrated. Flow rates through this valve were obtained by measuring the change in inventory in vessel A on occasions when there was no influx to this vessel. The level in vessel A was measured with a strain-gage-type transducer capable of easily sensing changes as small as 1/10 in. WC. (Records from three tests are shown in Figure 4-7.) The dimensions of Vessel A were such that a change of 1 inch in level represented a volume of 600 gallons. Data are presented in Table 4-2.

Table 4-2. Data for Calibration of Control Valve

Valve Opening, %	Pressure Drop, psi	Level Change, in./min	Flow Rate, gpm
16	58.4	.252	151
37.6	51.6	.379	227
56	39.8	.458	275
72	25.5	.530	318
84	16.1	.589	353
92.8	10.8	.712	427
100	6.0	.663	398
100	6.0	.653	392

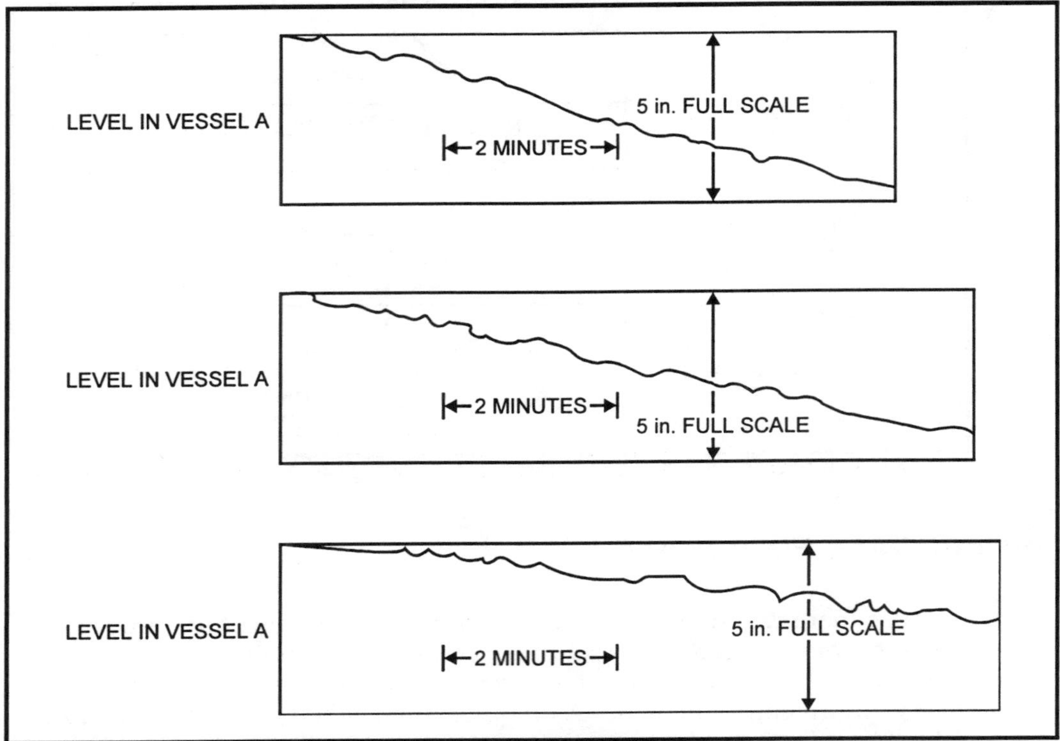

Figure 4-7. Level changes in vessel A when calibrating control valve

As has been shown elsewhere in these pages, the performance of valves of widely different design can be represented by the form

$$Q = Ke^{kz}\sqrt{\frac{\Delta P_v}{\rho}}$$

where

Q	=	volumetric flow rate
ΔP_v	=	pressure drop across valve
ρ	=	density of flowing fluid
z	=	position of valve stem, % open
K and k	=	are empirical constants peculiar to the valve

Data applicable to the above valve, given in Table 4-2, when plotted on semi-log coordinates, are shown in Figure 4-8. Since, in this case, the density of the liquid is constant, the relationship (derived from Figure 4-8) becomes

$$Q = 13e^{.0226z}\sqrt{\Delta P_v}$$

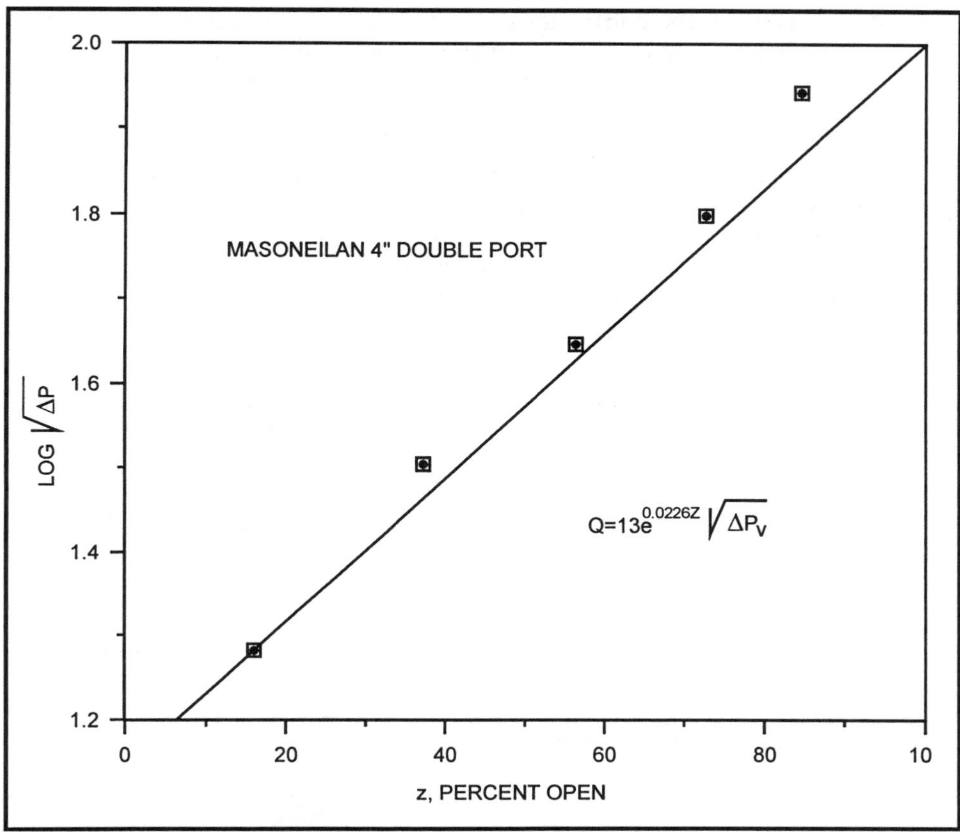

Figure 4-8. Calibration of Masonelian control valve

Hysteresis in the valve stem motion of this valve was less than 2% at any valve stem position, and the relation between the pressure to the positioner and that on the diaphragm was linear throughout the range of interest.

Several examples are given in this book where empirical relationships involving valve stem position, pressure gradient, and fluid density can be related to the rate of flow through valves. The purpose has been to emphasize that valves can frequently be used as reliable flow meters. In fact a well-designed valve, appropriately sized and properly located in a line, may well be superior to the usual orifice installation for metering both liquids and gases.

Operating with Level Controller in Manual Mode

The records shown in Figure 4-5 demonstrated that the controller regulating the flow rate of fresh feed, through the level control valve actuated by the level controller on vessel D, was contributing to the oscillatory behavior. This behavior suggested that the integral action imparted by the controller was either poorly adjusted or perhaps was unnecessary. When this controller was placed in the manual mode, the oscillations of many variables immediately disappeared. Figure 4-9 shows the result of deliberate changes in the position of the control valve (upper record). The ramp-like change in the level in vessel D clearly verifies that the system itself is an integrator. Following the step increase in the valve position (shown at point A in the top record of Figure 4-9), the level in vessel D increased at the rate of 0.44 in. WC/min. The increased flow rate through the valve was about 60 gpm occurring over a period of 3.5 minutes. The accumulation of 210 gallons was distributed in vessels B, C, and D, causing an increase of level in the latter vessel of 0.54 in. In a

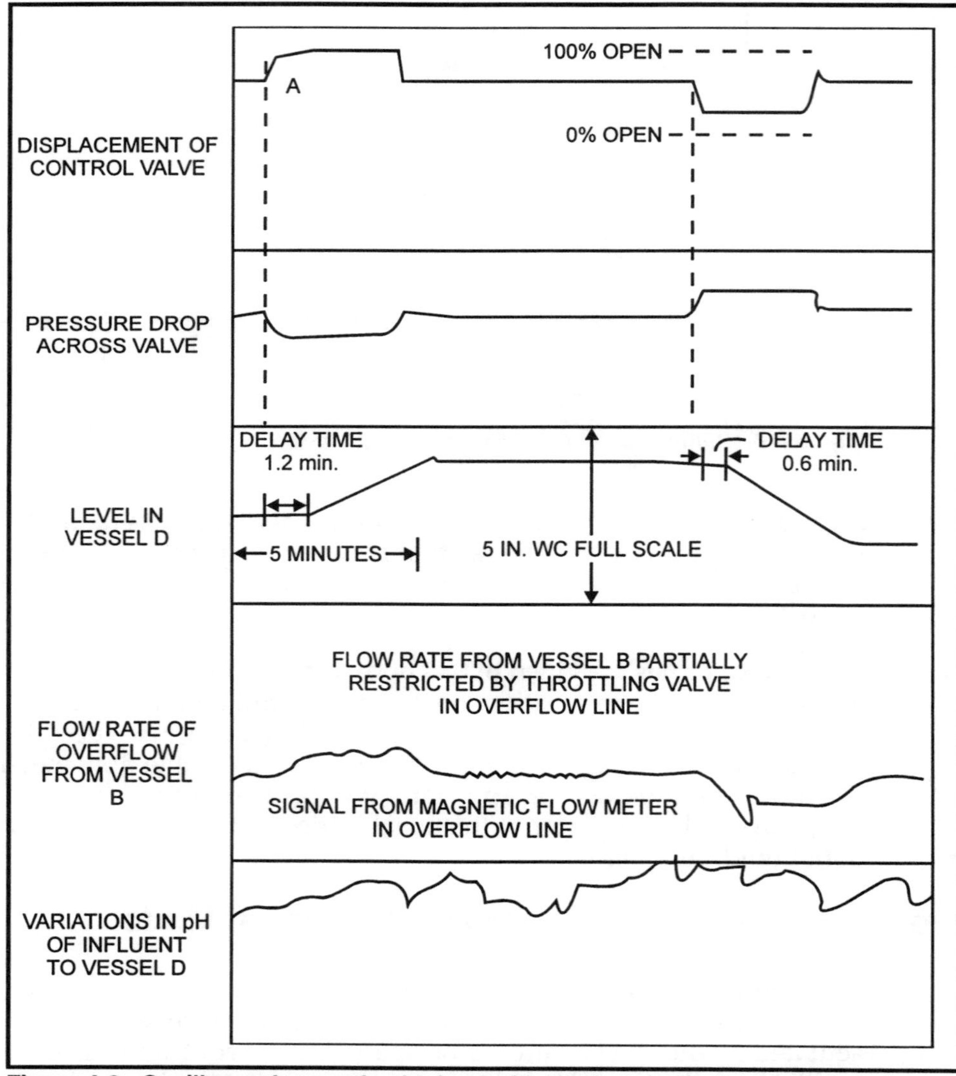

Figure 4-9. Oscillograph records obtained with level controller in manual mode

second test the feed rate was decreased from 314 to 152 gpm for 3.25 minutes. This resulted in a decrease in inventory of 527 gallons, causing a diminution of 0.75 in. in the level in vessel D. This demonstrated that the original controller adjustments were, in part, responsible for the observed oscillations in flow rate through the control valve.

Effect of Throttling Effluent from Vessel B

During a brief period of operation, a valve in the vertical line leading from vessel B to vessel C was partially closed. Apparently, by restricting the flow rate of overflow from vessel B, air entrainment was reduced, and the magnitude of flow changes became less severe. (This is shown in the second-to-lowest record of Figure 4-9.)

System Responses with Controllers in the Manual Mode

The third record from the top in Figure 4-9 shows the response of level in vessel D as the control valve in the feed line is manually moved from one position to another. The ramp-like response of level to step changes in flow rate through the valve verifies that the process is an integrator.

From the records shown in Figure 4-9, the delay time between a change in flow rate through the control valve and a change in level of vessel D can be obtained. These show that the delay time, following an increase in flow rate, is about 1.2 minutes.

Dynamic Response of Control Valve

The dynamics of the control valve were also determined. The pulse method of excitation was used, and the data reduced to frequency response by the trapezoidal approximation of the Fourier transform. Time histories of responses to typical positive and negative inputs are shown in Figure 4-10, along with the derived frequency response. First-order characteristics, with a time constant between 0.3 and 0.4 seconds, appear to apply.

Recommended Remedial Action

At this point the experimental work was concluded since, until the spurious changes in the rate of overflow from vessel B were eliminated, further work was not warranted. To eliminate this problem the arrangement indicated in Figure 4-11 was recommended. The objective was to reduce inclusion of air in the overflow from vessel B. Assuming a terminal velocity of air in water of 0.2 ft./sec and a liquid flow rate of 600 gpm, a cylindrical stand pipe with a diameter of 3 feet would suffice.

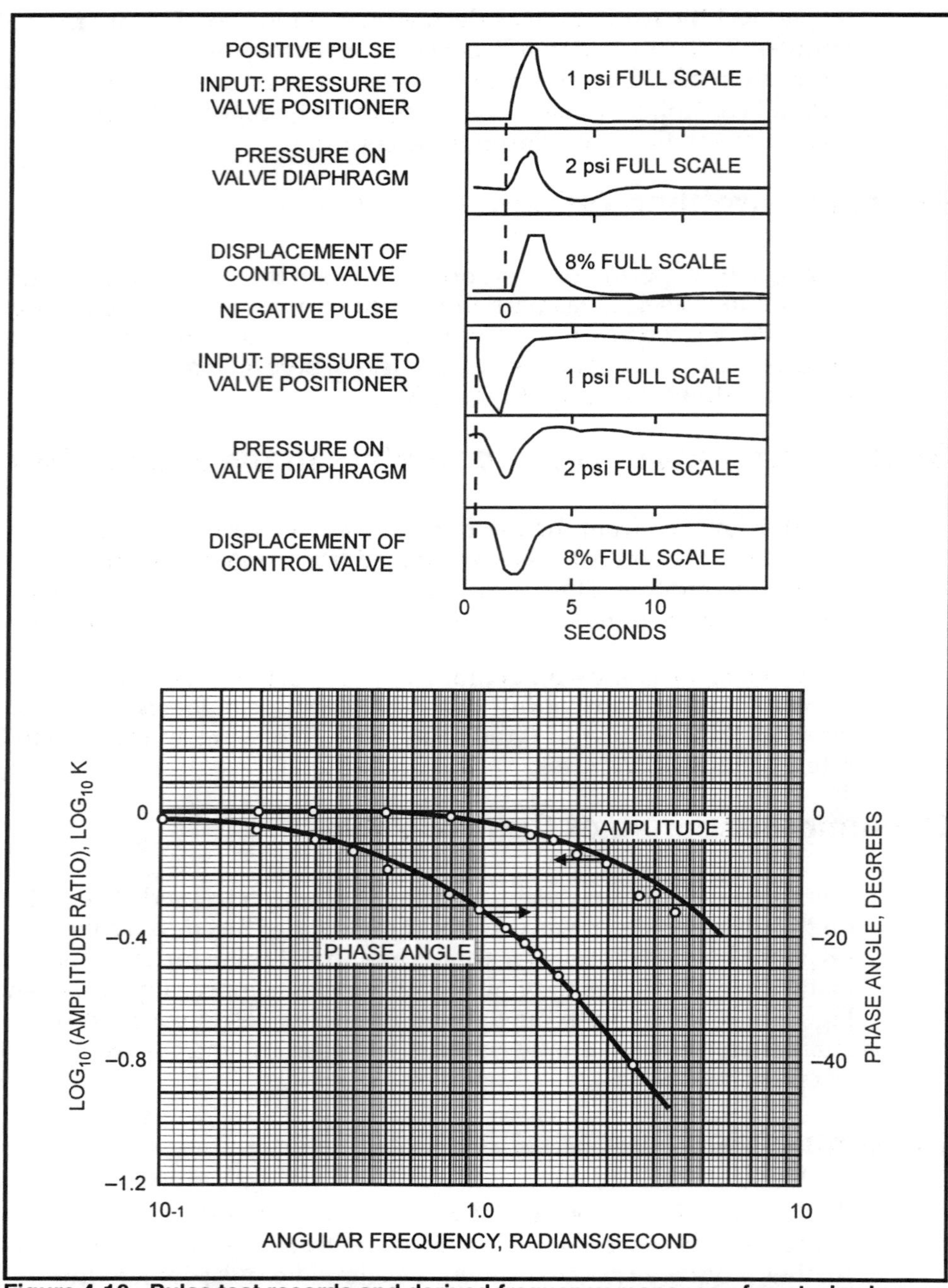

Figure 4-10. Pulse test records and derived frequency response of control valve

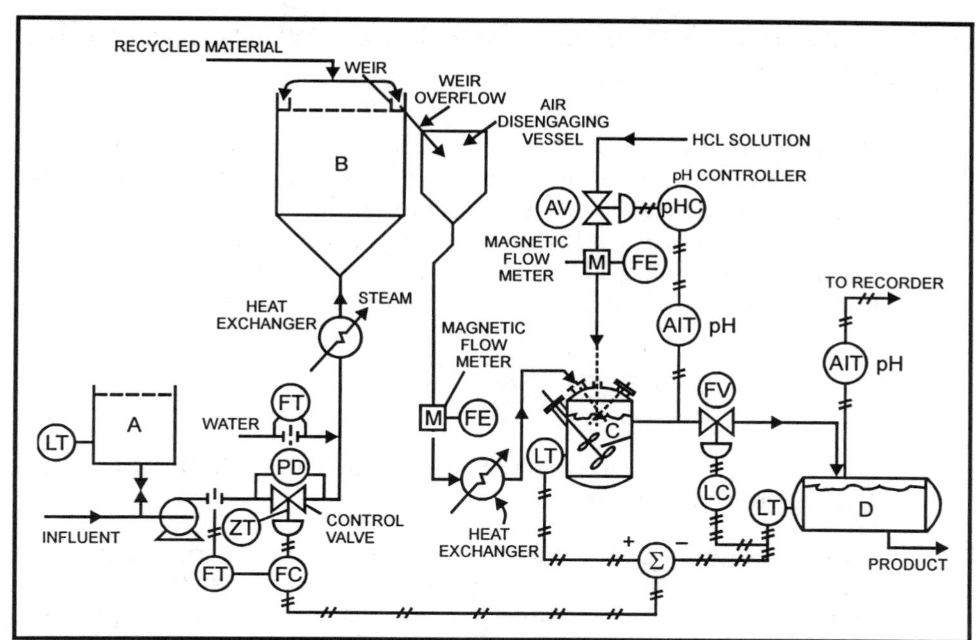

Figure 4-11. Process and instrumentation diagram with recommended control strategy

Suggested Control Strategy

The two principal objectives are to control the pH from, and liquid level in, vessel D. However, in order to eliminate the delay time between vessel C to D, controlling the pH of the effluent of vessel C was recommended. This is the vessel to which acid is added and is well agitated so that the dispersion of HCl should be rapid and uniform. The pH at this point, as well as that at vessel D, should be recorded and conveniently displayed. In view of the rather poor dynamics of the pH-sensing systems described here, precise control would be difficult to achieve.

Since the level in vessel D must be maintained very precisely, the strategy of controlling the difference in levels between vessels C and D might have merit, assuming that variations in the overflow from vessel B could be largely eliminated. The scheme suggests letting this difference regulate the flow rate of fresh material entering the process. Some reduction in pure delay time could also result from this arrangement. Therefore, metering and controlling the flow rate of the influent was recommended. Purely proportional control action, with a reasonably high gain, was also recommended for both controllers. Valve dynamics should be satisfactory, if of the quality shown by Figure 4-10. Figure 4-11 shows the proposed processing and control arrangement.

Using Integral (Reset) Control Action when Controlling Integrating Processes

Processes that possess integration properties exhibit a ramp response to a step input. This behavior has been shown to occur in this study: the level in vessel D increased in a ramp fashion following an abrupt increase in flow rate to the vessel. If we use a proportional + integral, *i.e.*, a P+I controller, two integrations appear in the feedforward path. Thus the feedforward function becomes, in the operator domain,

$$K_c \frac{(1 + T_i s)}{(T_i s)} \left(\frac{K_p}{s} \right),$$

where K_c and T_i are the controller gain and integration parameter, and K_p/s represents the process.

Thus, at zero frequency, the open-loop frequency response exhibits a phase angle of –180 degrees. If pure delay time exists the phase angle can easily be more negative than –180 degrees. In any case, compensating for a phase lag at higher frequencies becomes almost impossible. The above situation existed on the process described here and has been observed on numerous occasions.

Conclusions

In retrospect, at least some of the many difficulties mentioned in this study could have been predicted, or discovered, without undue effort. For example, the performance of pH-sensing and measuring instrumentation should have been available either from vendors or from research laboratories. In addition, the detrimental effects of delay time between sensors and points of corrective action were rather obvious. Finally, the rather large variations in the overflow from vessel B should have been expected and diagnosed prior to and certainly early in this study. Hind sight, however, is always more widely possessed than foresight!

What Was Learned from This Study

- Always suspect unsteady flow rates when liquids flow vertically by gravity from open vessels through closed conduits. And, as a cardinal rule, never install flow meters in such conduits!
- The exceptionally nonlinear response of pH sensors always imposes special problems, if these sensors are part of a closed-loop feedback control system.
- The preparation for experimental studies of processes should invariably include inspection of every item of importance, as well as the verification of their performance.

- Controller functions should be chosen with some consideration of the dynamics of the process to be controlled. Actual dynamic information, although desirable, is not always required to avoid misapplications.
- Properly calibrated control valves can sometimes be used as reliable flow meters.
- Improperly selected controller parameters can be the cause of sustained disturbances.

Finale

The test crew was denied the privilege of knowing if the recommended changes were implemented, since no feedback was forthcoming from the client. Perhaps this indicated satisfaction with the results derived from this study.

5

Optimizing Maleic Anhydride Production

An experimental study of the process by which maleic anhydride (MA) is produced is presented here. This work was completed many years ago; in the meantime, the choice of raw material and catalysts have changed, as well as many details of the processing scheme. Nonetheless, the work described here serves to demonstrate that comprehensive and careful experimental studies of operating processes can produce useful results, leading to improved economics.

The request for conducting these tests came from the production department, after previous work of considerable magnitude had not produced entirely satisfactory results. In this previous work an attempt was made to apply statistically designed experiments, in which independent variables were changed in a prescribed manner, in the hope of developing meaningful relationships among production rate, quality, and costs. In defense of that work, it should be stated that most of the plant instrumentation used in those studies produced data from which there was little hope of retrieving valid information of any kind. Accuracy and sensitivity were the chief limitations of the existing instruments. While most of this prior work was of little value to the next group of investigators, it emphasized the need for better instrumentation and served to indicate where improvements in data collection were imperative. Consequently, proven test instrumentation was installed to procure all vital information, quite independent of existing plant instrumentation. In addition, sensors for some variables not previously measured were added.

Perhaps the most useful information obtained as a result of the improved data acquisition system were rates of energy developed in the reactor, heat transfer to the molten salt medium, and energy rejection in generated steam. Energy and material balances around some processing components could be closed to within 2% (this strict accounting of energy transfer being crucial to the success of this study).

The Reactions

The ideal partial oxidation of benzene to maleic anhydride is shown below,

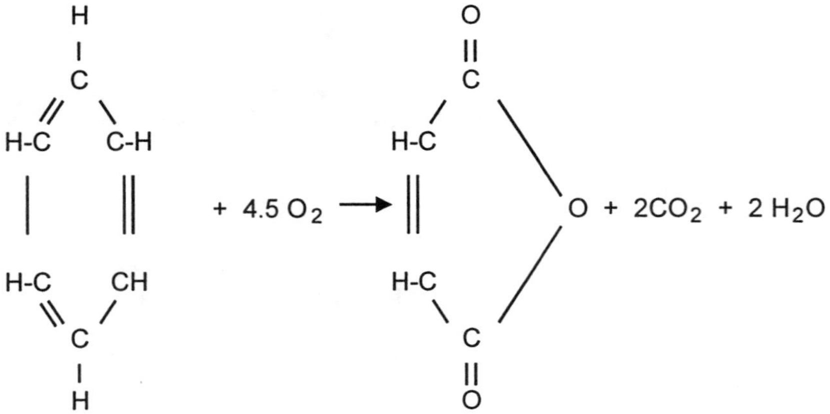

```
         H                                  O
         |                                  ||
         C                                  C
       //  \                              /    \
   H-C       C-H                      H-C        \
    |        ||                        ||          O + 2CO2 + 2 H2O
    |        ||      + 4.5 O2  --->    ||        /
   H-C       CH                       H-C       /
      \\    /                            \     /
         C                                  C
         |                                  ||
         H                                  O
```

Competing with this reaction is the complete oxidation of benzene:

$$C_6H_6 + 7^1/_2\,O_2 \longrightarrow 6CO_2 + 3\,H_2O$$

At intermediate stages CO also appears.

The objective is to establish conditions within the reactor such that the first reaction is enhanced with a minimum of the other reactions.

The Process

The reactor consisted of a bundle of vertical tubes filled with a granular catalyst. The removal of thermal energy from the exothermic reaction was accomplished by circulating liquid heat transfer medium, a molten mixture of inorganic salts, through the tube bundle. (Figure 5-1 shows one processing arrangement.)

The usual oxidant is air into which the hydrocarbon (in this case benzene) was aspirated. The mixture, preheated by heat exchange with reactor effluent, was admitted to the lower end of the reactor. Reactor effluent was cooled in a steam-generating exchanger, after which it served to preheat the incoming air-benzene mixture. Finally this stream passed through a water-cooled condenser, the after- cooler, in which the majority of maleic anhydride was condensed. The temperature of water to this condenser was maintained high enough to avoid deposition of solid maleic anhydride on the condensing surfaces.

Uncondensed maleic anhydride from the partial condenser was combined with water to yield maleic acid, this being later recovered and dehydrated to obtain additional maleic anhydride. This study did not encompass the recovery system.

As mentioned above, thermal energy is removed from the reactor by a circulating heat transfer medium. This is subsequently cooled in a steam-generating exchanger and recycled through the reactor by a pump immersed in the salt sump. The heat transfer medium was a proprietary mixture of various salts that remained molten at the prevailing temperatures. Reactor effluent was cooled in another steam generator after which it was used to preheat the incoming mixture of air and benzene. Final cooling of the reacted material was accomplished in the after-cooler, where the majority of the maleic anhydride produced was condensed; it was then pumped to storage.

Existing Instrumentation

The existing sensing and control systems are shown in Figure 5-1. The performance of these control systems left much to be desired, and manual intervention was frequently exercised. Manual control of the flow rate of the coolant to the reactor was almost continuous since positioning of this valve was particularly difficult. The environment in which this molten salt flow-regulating valve operated was exceedingly severe.

The Test Measuring System

At the time of this study the state of industrial instrumentation was such that the probability of obtaining definitive data from existing plant sensors was virtually zero. Fortunately this was recognized, and adequate instrumentation obtained and installed.

The points at which sensors were placed are shown in the diagram of the process, Figure 5-1. All temperature and temperature differences were measured with calibrated resistance elements having bridge outputs of one millivolt per degree F. The strain gage transducers used to measure pressure and pressure differential were excited with 10 volts dc and had full range outputs of 5 millivolts per volt of excitation. Flow rates of make-up water to the steam drum, liquid maleic anhydride, and benzene were measured with calibrated turbine flow meters, the flow rate of process air with a Dall tube, and cooling water to each of the steam generators with factory calibrated flow nozzles. Differential pressures developed by these latter sensors were measured with differential pressure strain-gage transducers. The flow rate of the molten salt heat transfer fluid was metered with a flow loop. This device was shop fabricated from a section of 3-inch stainless steel pipe and shaped into a circular loop with an internal diameter of about 10 inches. The leads from the molten salt were electrically traced to avoid solidification in the interconnecting tubing and reservoirs. A flexible metal diaphragm separated the molten salt from the

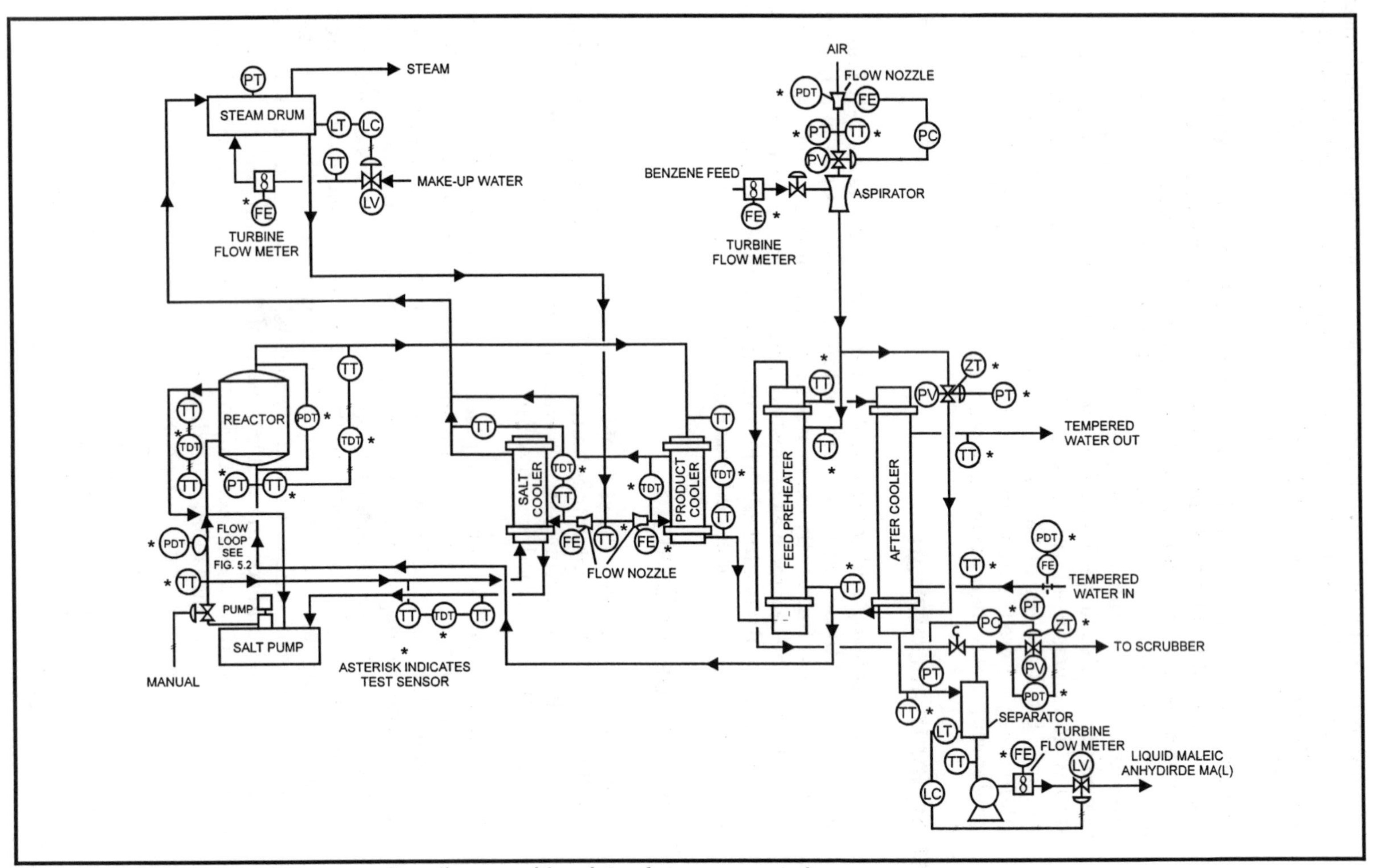

Figure 5-1. Maleic anhydride process flow diagram with points of measurement shown

oil to which the transducer was exposed. (The flow-loop assembly is shown in Figure 5-2.)

Since the pressure gradient (which is developed across the taps located diametrically and horizontally across from one another in a circular section of the loop) is dependent only on the rate of momentum change at that position, compensation for viscosity is not required. The signal developed becomes a measure of the fluid velocity and density and, hence, was calibrated using water. Throughout the testing program, this meter performed satisfactorily and enabled computation of heat removal rates from the reactor, as well as from the salt cooler, with great precision.

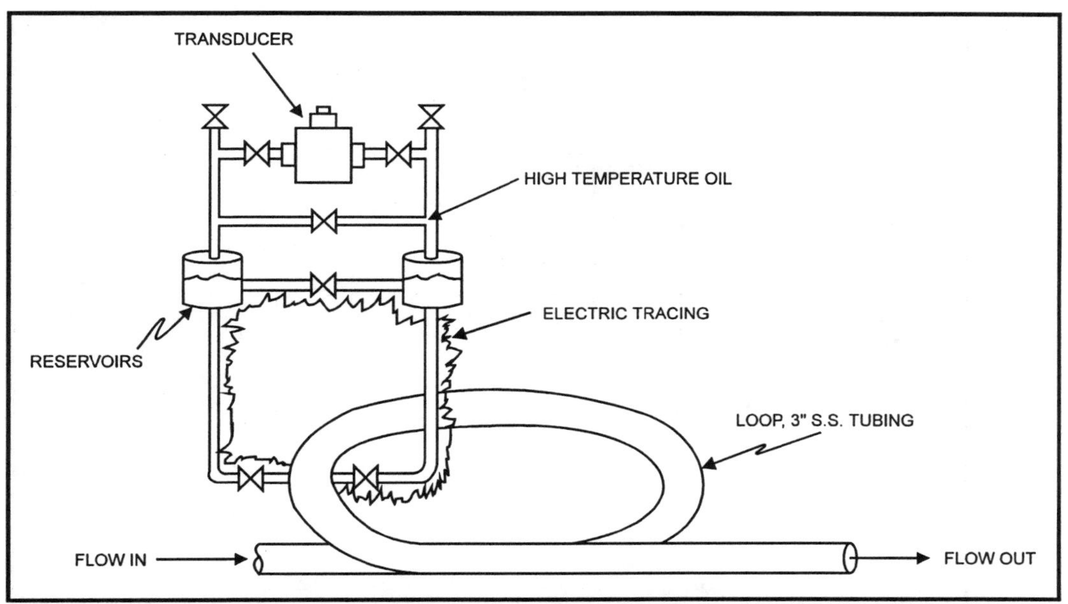

Figure 5-2. Flow loop used to measure the flow rate of molten heat transfer medium

Pressures to all valve positioners and valve diaphragms, as well as valve stem position, were measured to determine the presence of hysteresis, "stiction," and other malfunctions of these components.

Transducer outputs were directed to multichannel oscillographs provided with high-impedance, high-gain dc amplifiers capable of sensing a signal change as small as 2 microvolts. Calibrated signal suppression enabled nulling out any amount of the signal so that very small perturbations around the steady state of any signal could be amplified, observed, and recorded. Adjustable filters on the input to each amplifier enabled removal of high frequency signal corruption that might appear. Chart speed could be varied over a considerable range which permitted recording rapidly varying signals, including responses to any step- or pulse-like signals deliberately introduced. All data recording instrumentation was located in a separate building, somewhat removed from the processing facilities.

Transducer signals were transmitted through shielded cables with a common ground.

Compositions of reactor effluent and noncondensed material from the after- cooler condenser were obtained from analyzers normally associated with the plant. Great care was taken to ensure that valid samples were obtained and the instruments calibrated. Figure 5-3 shows the arrangement of the sampling system and the instrumentation used to obtain the composition of the effluent from the after-cooler. Analyses of samples taken from other points in the process were obtained in a similar manner with suitable instruments.

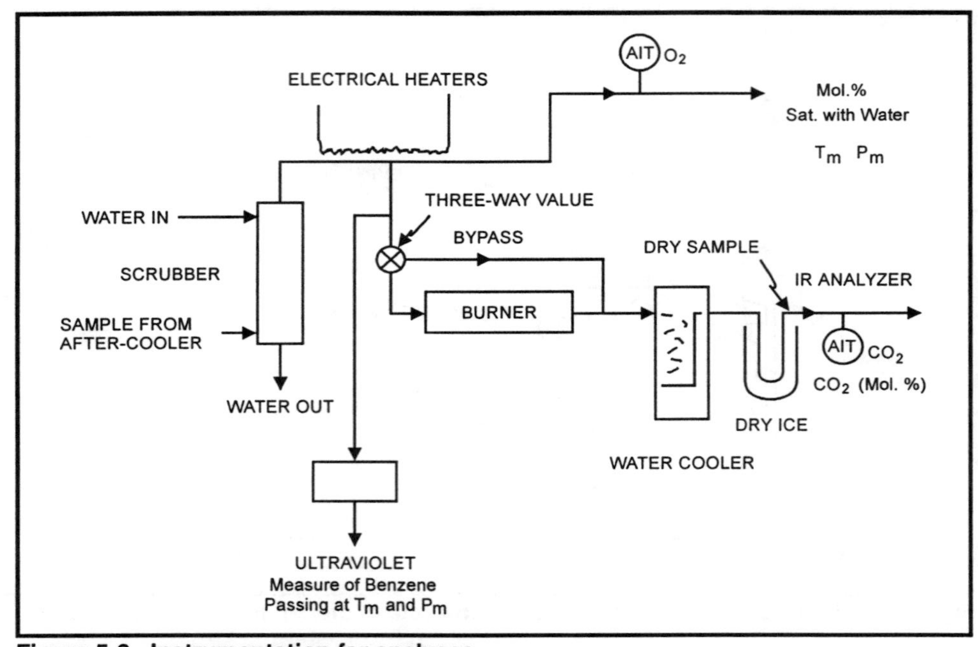

Figure 5-3. Instrumentation for analyses

The Testing Program

Data collection was designed to satisfy both statistical and classical methods. To meet the former required more time than if only the latter method were used in this study. Because of the small number of independent variables, the classical technique of varying only one variable at a time proved to be the better approach. The data collection was completed in 6 weeks, which could have been sharply reduced were it not necessary to accommodate the smoking and shift breaks required by plant operating personnel.

An inordinate amount of effort and time was spent recording and reducing signals from a number of thermocouples located in several reactor tubes at various elevations. These were intended to locate regions of unusually high temperatures, if such existed. No useful information was retrieved.

The classical method of experimental investigations has two distinct advantages. One is that the investigator is required to work closely with the real system and not do his thinking "through a mathematical model." He cannot isolate himself from reality. The other advantage is that data, which appear erratic or "out of line," usually become obvious. All steps producing the result in question can be scrutinized for possible errors in recording, data reduction, or uniqueness in respect to experimental conditions. Equal weight is not given to all the information unless data relevant to that information are consistent and trustworthy. Knowledge of the process and the data system is utilized, and judicious discrimination is permitted in the evaluation of the information.

Some dynamic response data were obtained. These were of special interest to the test crew and may represent the first of this kind of information obtained on a reactor of this design and, of course, apply only to this particular plant. A reactor of different configuration would yield other relationships.

No data were obtained specifically to elucidate kinetic mechanisms, nor are any speculations of such offered. Whereas the plant is the place to test kinetic theories, it is certainly no place to develop the theory of fundamental kinetic mechanisms. This must be done in the laboratory in a carefully controlled environment where extremely accurate measurements can be made of catalyst surface conditions within a reactor.

Material and Energy Balances

The measurements and calculations leading up to material and energy balances were essential for procuring, substantiating, and producing the results achieved in this study. For this reason, the most important calculations associated with the reactor are indicated (as shown in Figure 5-4).

Material Balances

Input items included oxygen, nitrogen, and water vapor entering with the air and, of course, the benzene added and the incoming molten salt. The portion of components that enter and leave unreacted are treated separately. This includes unreacted benzene and oxygen, nitrogen and water vapor.

Energy Balance Around the Reactor

The temperature difference, ΔT_1, applies to all material entering and leaving the catalyst bed. Effluent material also carries energy produced by the reaction, the molten salt, of course, absorbing the majority of the latter.

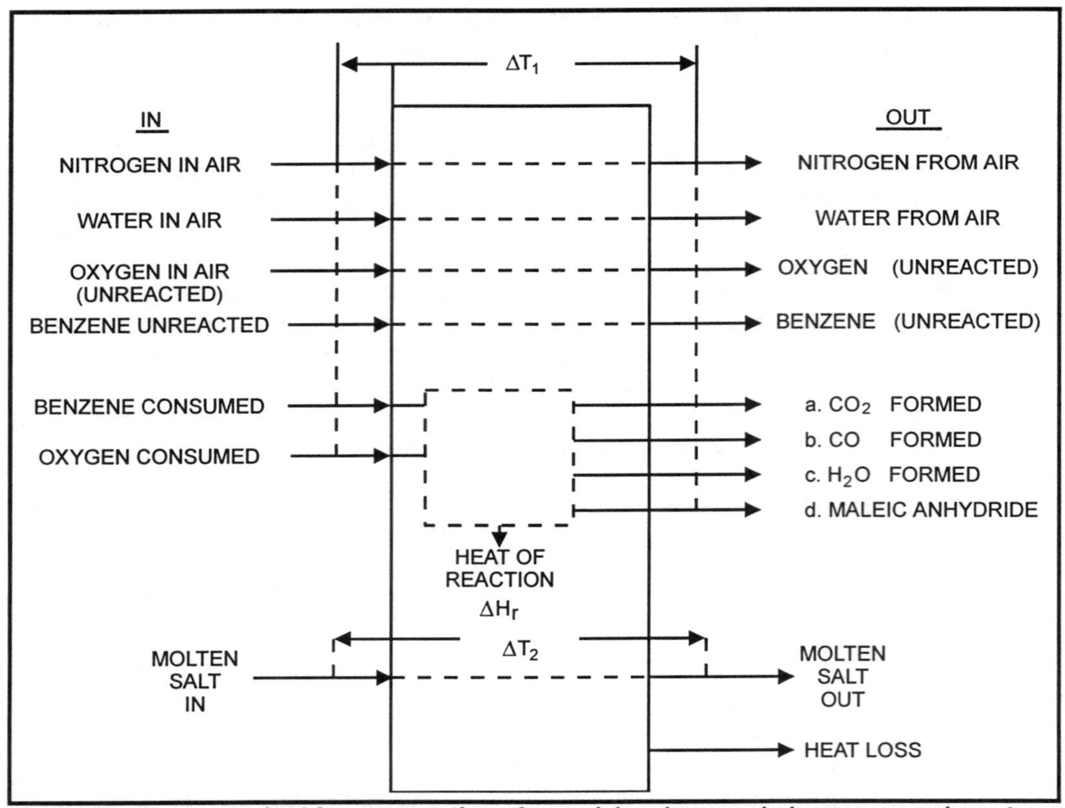

Figure 5-4. Items required for computation of material and energy balances around reactor

The temperature difference, shown as ΔT_2, is the temperature rise experienced by the heat transfer medium.

Other Energy and Material Balances

The relationships needed to obtain energy and material balances around the other heat exchangers are not shown. Of particular interest were the after-cooler and the two steam generators. In the latter devices about 38% of the water was consistently converted to steam.

The painstaking derivation of all the relationships needed to recover the desired information are not given here; an inordinate amount of space would be required. Suffice it that this task was no trivial exercise. However, a list of several energy relations is shown from which important information was obtained.

Referring to Figure 5-4, the heat of reaction appears in the relation:

$$Q_p - Q_r + Q_{ar} = \{\Delta H_r\,(25°) - \text{Losses}\}$$

where

Q_p = the enthalpy of the products above 25°C
Q_r = the enthalpy of reactants above 25°C
ΔH_r = the heat of reaction at 25°C, and
Q_{ar} = $Q_{ugr} + Q_{cr}$, where the enthalpy in the unreacted gases is
Q_{ugr} = $Q_{nr} + Q_{or} + Q_{wr} + Q_{bpr}$, and
Q_{cr} = heat absorbed by coolant
Q_{nr} = enthalpy of nitrogen
Q_{or} = enthalpy of oxygen
Q_{wr} = enthalpy in water, and
Q_{bpr} = enthalpy in benzene passing

Qualitative Considerations of Reactor Performance

Time histories of temperature along the reactor as the flow rate of feed increases, with constant inlet air to benzene ratio, are visualized in Figure 5-5. Each time history begins with a transient preheat period before reaching the "ignition," or "initializing," temperature where reaction begins. Profile A corresponds to a low flow rate of reactants. Here it is assumed that residence time within the reactor is excessive, which causes over-oxidation and a net loss of maleic anhydride. Reducing the flow rate of reactants (see profile B) enhances conversion to anhydride, since the time for competing reactions has been reduced. Time histories C and D are both assumed to be conditions where residence time is insufficient and too much benzene passes through unreacted. Thus, there exists an optimum residence time for a given reactor design, coolant flow rate, inlet temperature, and reactor pressure. In each case, as heat is absorbed, reactor coolant temperature increases, the magnitude of change being dependent upon the flow rate of coolant and rate of heat released within the reactor.

The experimental work described here was aimed, in part, at determining optimum operating conditions, subject to limitations of the existing processing apparatus.

Test Results—The Effect of Reactor Pressure

Reactor pressure could be varied either by throttling the inlet air or the effluent from the separator, the latter being preferred, although the reactor pressure then becomes largely dependent upon the pressure losses incurred in the process. The experimental results (given in Figure 5-6) show that reactor pressure could be changed only about 1 psi when throttling separator effluent. Air flow rate was changed from 9,000 to 11,000 scf/min. in these tests. The points within each cluster are data from observations at nearly the same conditions, conditions that are different for each group.

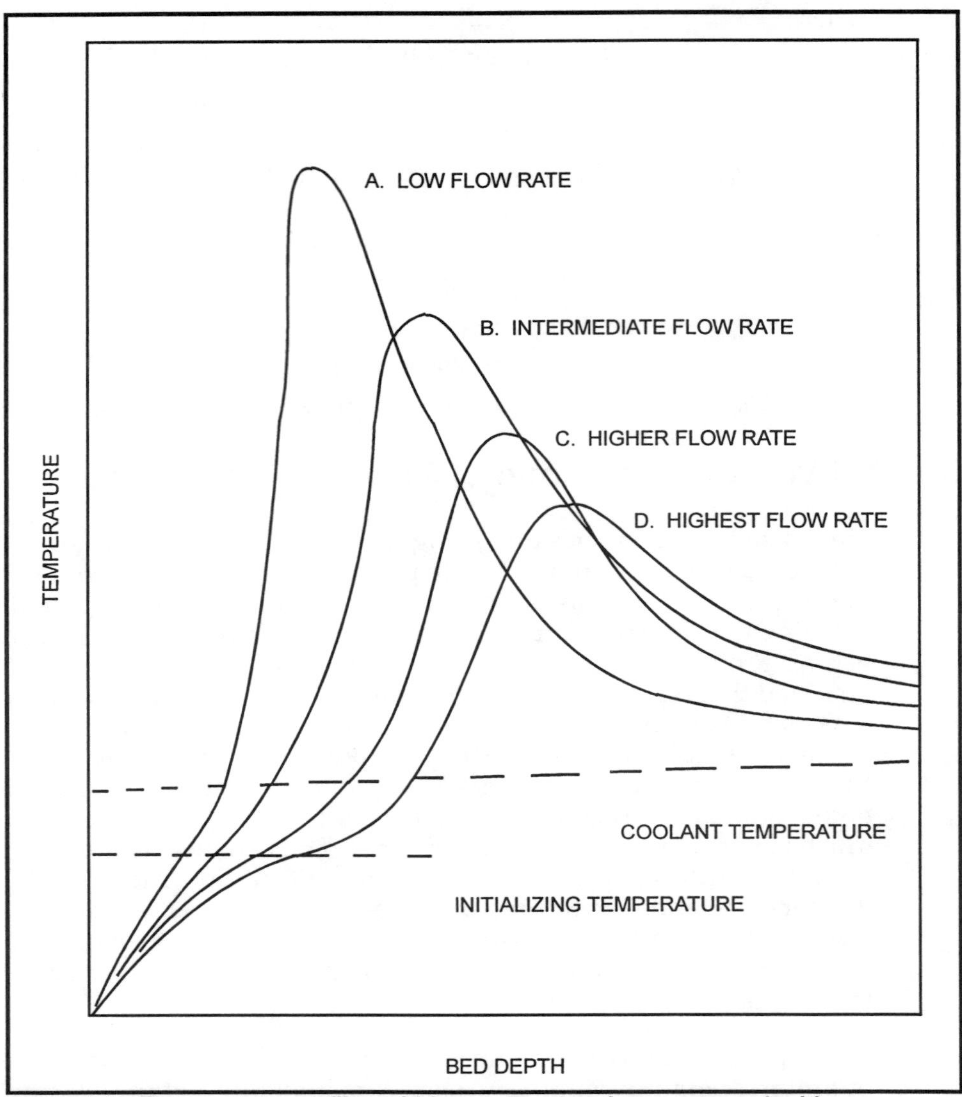

Figure 5-5. Temperature profiles within reactor at various space velocities (assumed)

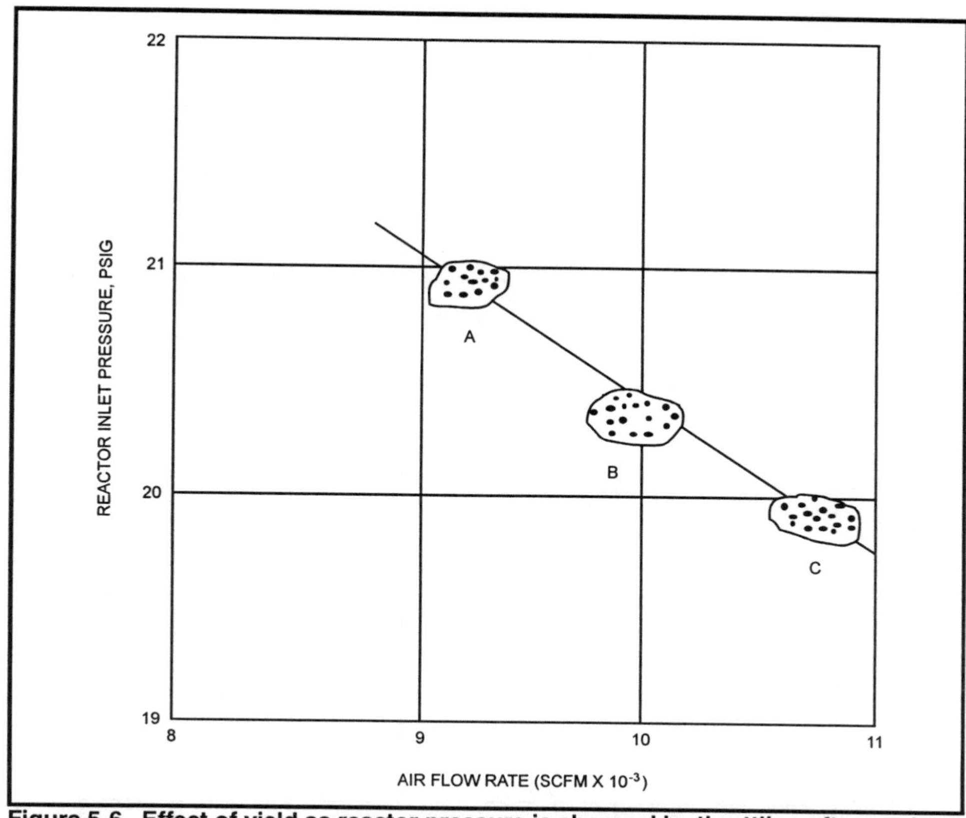

Figure 5-6. Effect of yield as reactor pressure is changed by throttling after-cooler effluent

Of course, by throttling the inlet air supply, greater diminutions in reactor pressure could be achieved. (This is shown by the lower curve in Figure 5-7.) An increase from 65 to 80 lb MA formed per 100 lb of benzene occurred by changing the reactor pressure from 15 to 22 psi. At the same time the ratio of maleic anhydride in the condenser overhead to that recovered in the liquid state decreases from about 0.45 to 0.3, as seen in the results shown in the upper part of Figure 5-7. Higher after-cooler pressure obviously increases condensation of product.

Effect of Flow Rate of Air and Reactor Temperature

Figure 5-8 demonstrates that conversion is favored at both low air flow rates and low inlet reactor temperatures, independent of air-to-benzene ratios between 350 and 440 scf of air per pound of benzene.

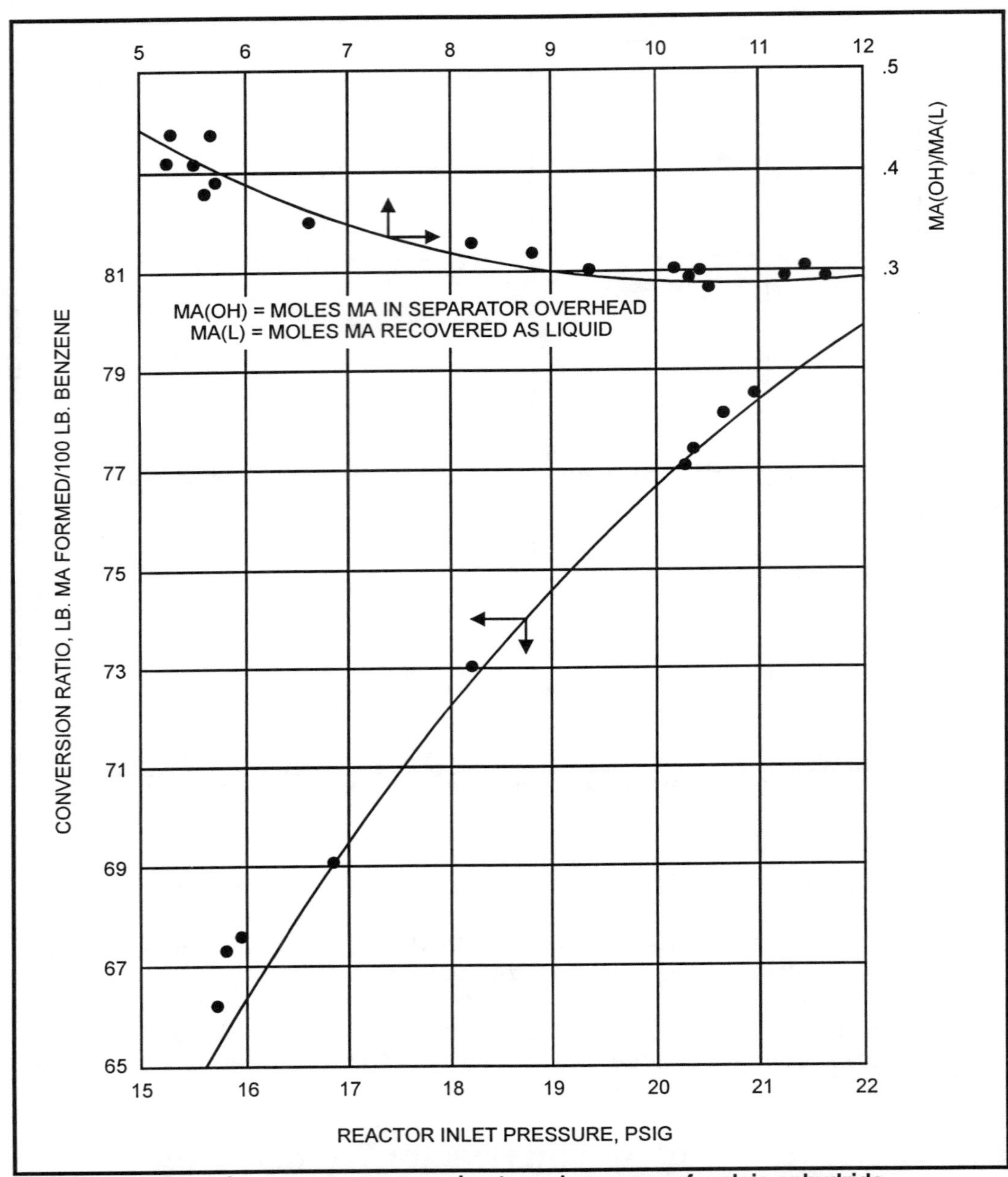

Figure 5-7. Effect of pressure on conversion to and recovery of maleic anhydride

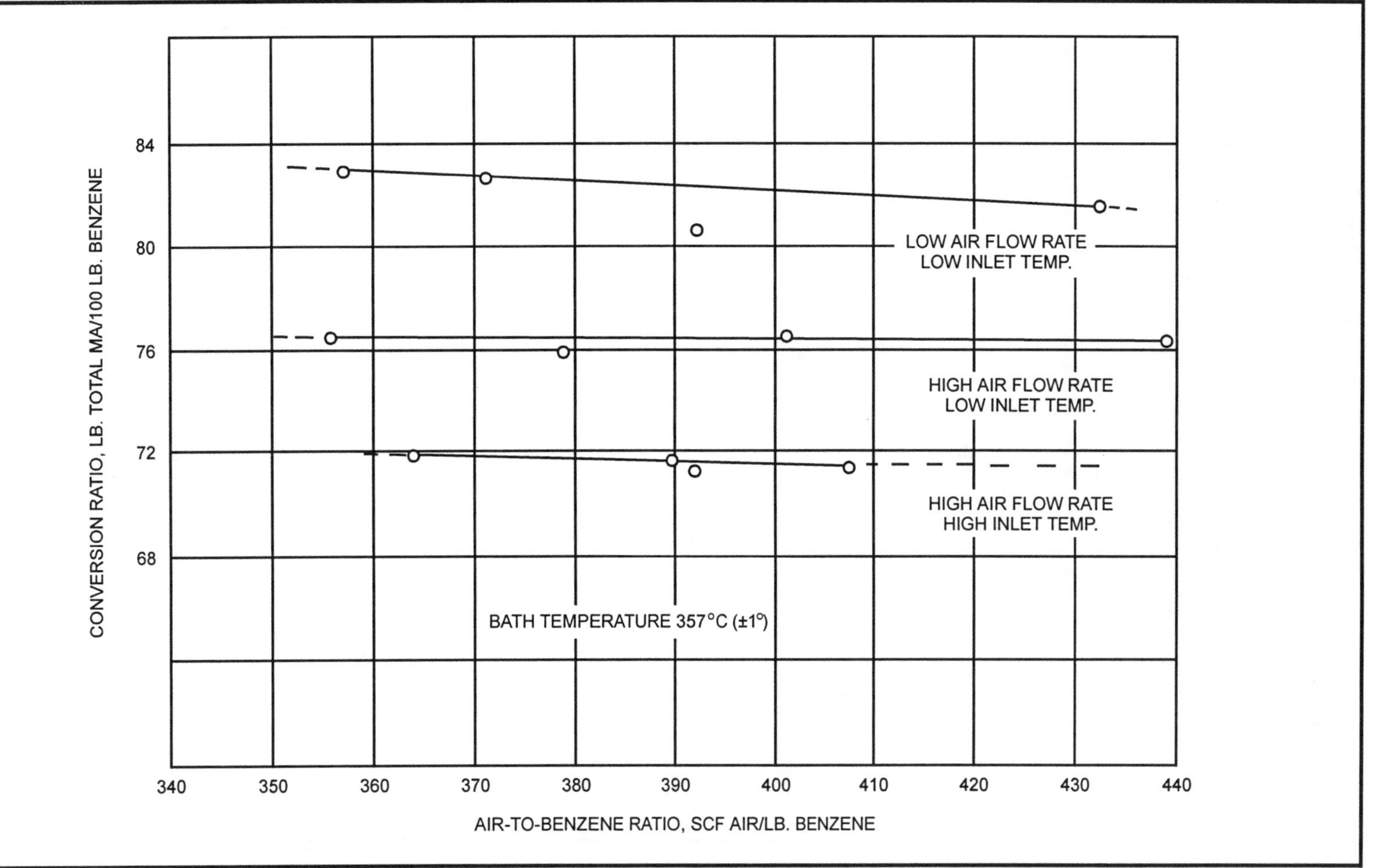

Figure 5-8. Effect of air-to-benzene ratio and reactor inlet temperature on conversion

Relation Between Total Maleic Anhydride Produced and That Recovered

Figure 5-9 clearly indicates, as expected, that the recovery of crude maleic anhydride liquid from the after-cooler is favored by low air rates and high rates of tempered water. The dashed line in the lower right-hand corner represents complete recovery as liquid product.

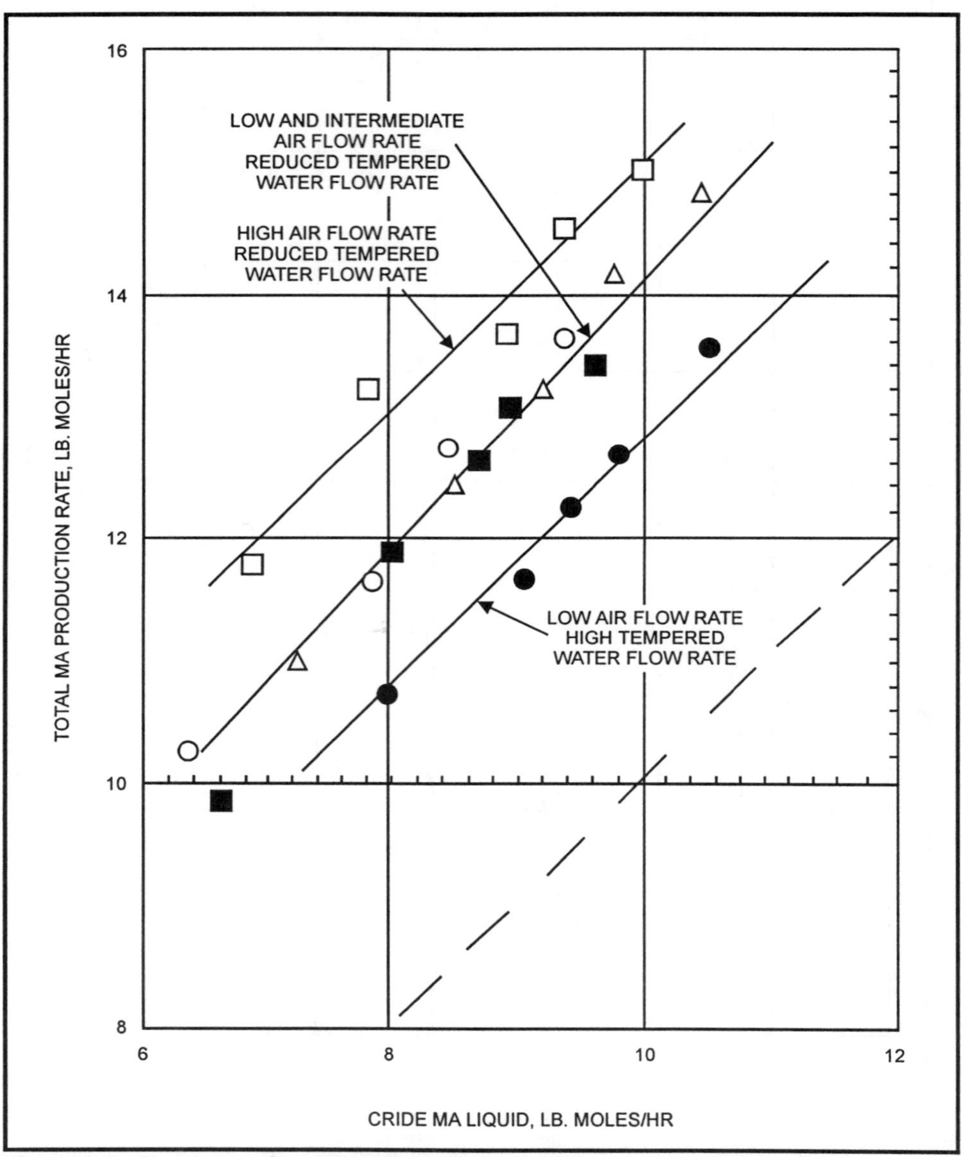

Figure 5-9. Recovery of maleic anhydride as liquid vs. total maleic anyhride produced at various air and tempered water flow rates through the after-cooler

An Approach to Optimization

The formation of maleic anhydride by the catalytic partial oxidation of benzene proceeds with the evolution of thermal energy. Competing with this reaction is the complete oxidation of benzene to carbon dioxide and water. The latter obviously releases more energy than the former per mole of benzene. Two extremes exist: if no heat is released there will be no maleic anhydride, or there can be maximum heat release with only carbon dioxide and water produced, again none of the desired product. Somewhere between these extremes lies the optimum production with a corresponding unique rate of heat release.

The rates of energy transfer in the various exchangers, computed from experimental measurements, showed that about 85% of the heat of reaction was consistently absorbed in producing steam. This figure was established from heat and material balances computed from the experimental data and compared to the theoretical heat released from the measured conversion. Confidence in these results was gained from the fact that energy and material balances around crucial parts of the process, notably the reactor, salt cooler, and steam generators, closed within 2%!

Since the flow rate of water to the make-up steam drum could be measured with great precision, and the enthalpy change between incoming water and effluent steam was almost constant, the flow rate of make-up water was almost proportional to the heat of reaction. The set of curves shown in Figure 5-10, corresponding to conditions of approximately constant cooling rates in the after-cooler, clearly indicate that high salt bath temperatures and low air flow rates favor the conversion of benzene to maleic anhydride in this reactor.

Figure 5-10 contains the information needed to develop a practical scheme to optimize conversion of benzene to maleic anhydride in this facility. First, it should be noted that maximum conversion occurs when the make-up water rate lies between 14,000 and 15,000 lb/hr. The important variables, for the facilities under study, were reactor inlet temperature, salt bath temperature, rate of air flow, and ratio of benzene to air. If reactor pressure could be manipulated independently of air supply rates, this also would have been a variable. In any event, for the system studied the procedure could be as follows:

- The minimum ratio of air to benzene is selected consistent with safety considerations.
- The flow rates of air and benzene are increased until the desired production is attained using operating conditions considered acceptable and approximately optimum. (Total production can be estimated from Figure 5-9.)

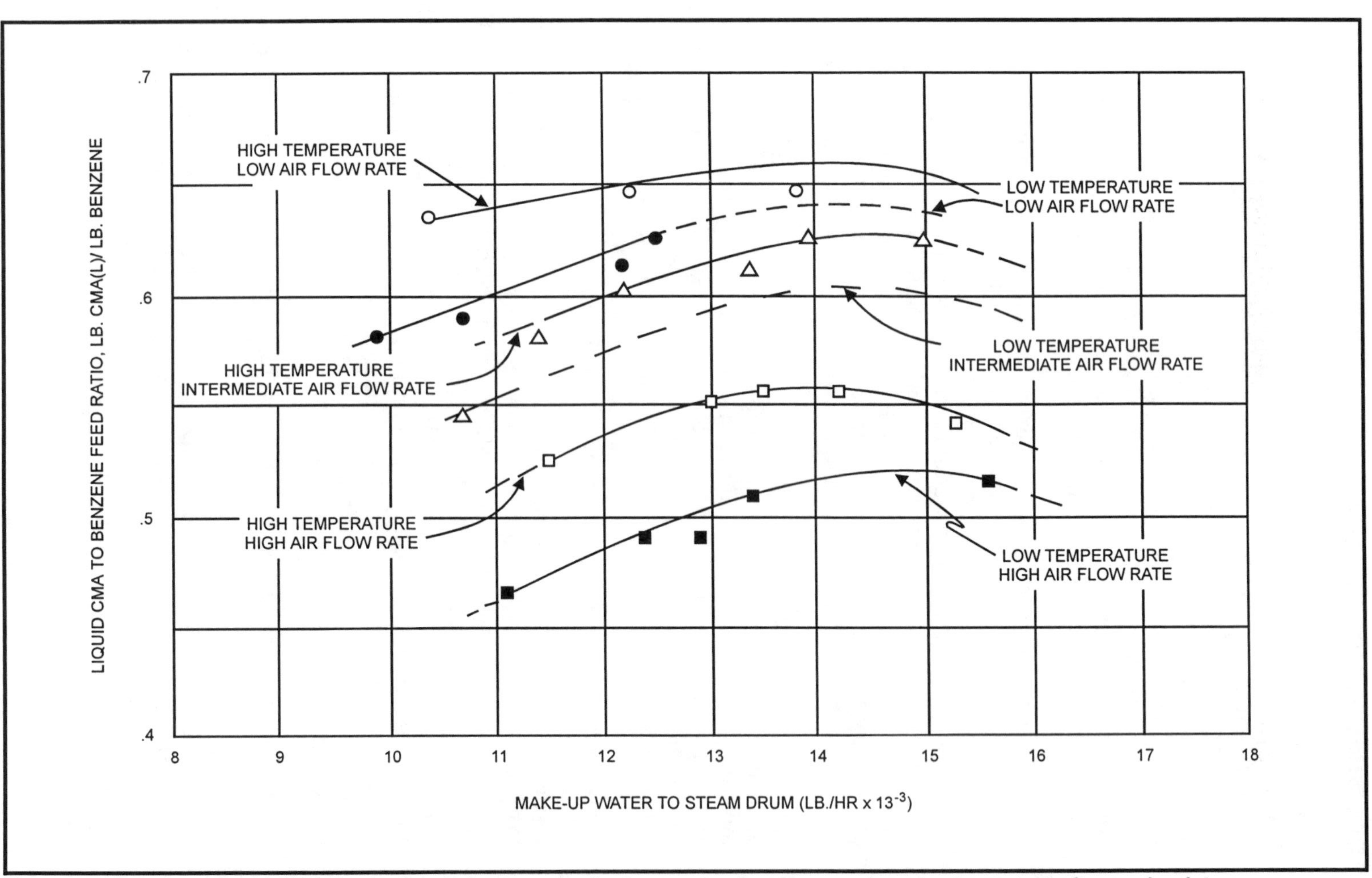

Figure 5-10. Ratio of liquid CMA production to benzene feed rate vs. rate of make-up water rate at constant after-cooler duty

- Adjustments in the independent variables are made in a systematic manner until the ratio of MA liquid production rate to benzene feed rate (*i.e.*, MA recovered as liquid/benzene) is maximum.

- If the make-up water flow rate is not between 14,000 and 15,000 lb/hr, the maximum ratio has not been achieved. A diminution in air flow rate and consequent reduction in air-to-benzene ratio probably will be required. However, this may be inconsistent with established safety procedures. In any event, comparing the actual make-up water rate to the proposed figure serves as a measure of the approach to minimum production costs per unit of maleic anhydride.

- Final adjustments of flow rates can now be made to ensure the desired overall production rate.

The simplicity of the measurements and suggested procedures make optimization both practical and easy. The three critical flow rates, benzene feed, liquid maleic anhydride, and make-up water can all be measured by turbine flow meters, for example, from which convenient, accurate, linear, easily-handled, millivolt signals are produced. For example, the ratio of the production rate of liquid maleic anhydride to benzene could be readily obtained as a single record. When presented along with the make-up water flow rate, a very convenient performance index could be continuously presented to plant operators. With today's data acquisition, handling, and computation capability, full automation of the optimizing procedure could be implemented.

Further Use of the Data

The experimental results may also be used to predict the maximum conversion ratio, as well as corresponding operating conditions at which this might be achieved, with the existing reactor and recovery system. From Figures 5-9 and 5-10 observe that both low air flow rates and high bath temperatures favor the conversion ratio as well as liquid production ratio. Figure 5-11 presents this information in another fashion. Here conversion ratio is plotted versus air flow rate for high and low temperature levels. Extrapolation of the constant temperature lines to a point of intersection suggests that a maximum conversion of about 86 lb of maleic anhydride per 100 lb of benzene might be achieved in this reactor, at an air flow rate of about 8,000 scf/m. This assumes no restrictions for reason of safety or acceptable after-cooler operating conditions.

With the information presented, optimum operating conditions can be established for the system studied herein, subject to the limitations imposed by existing plant features.

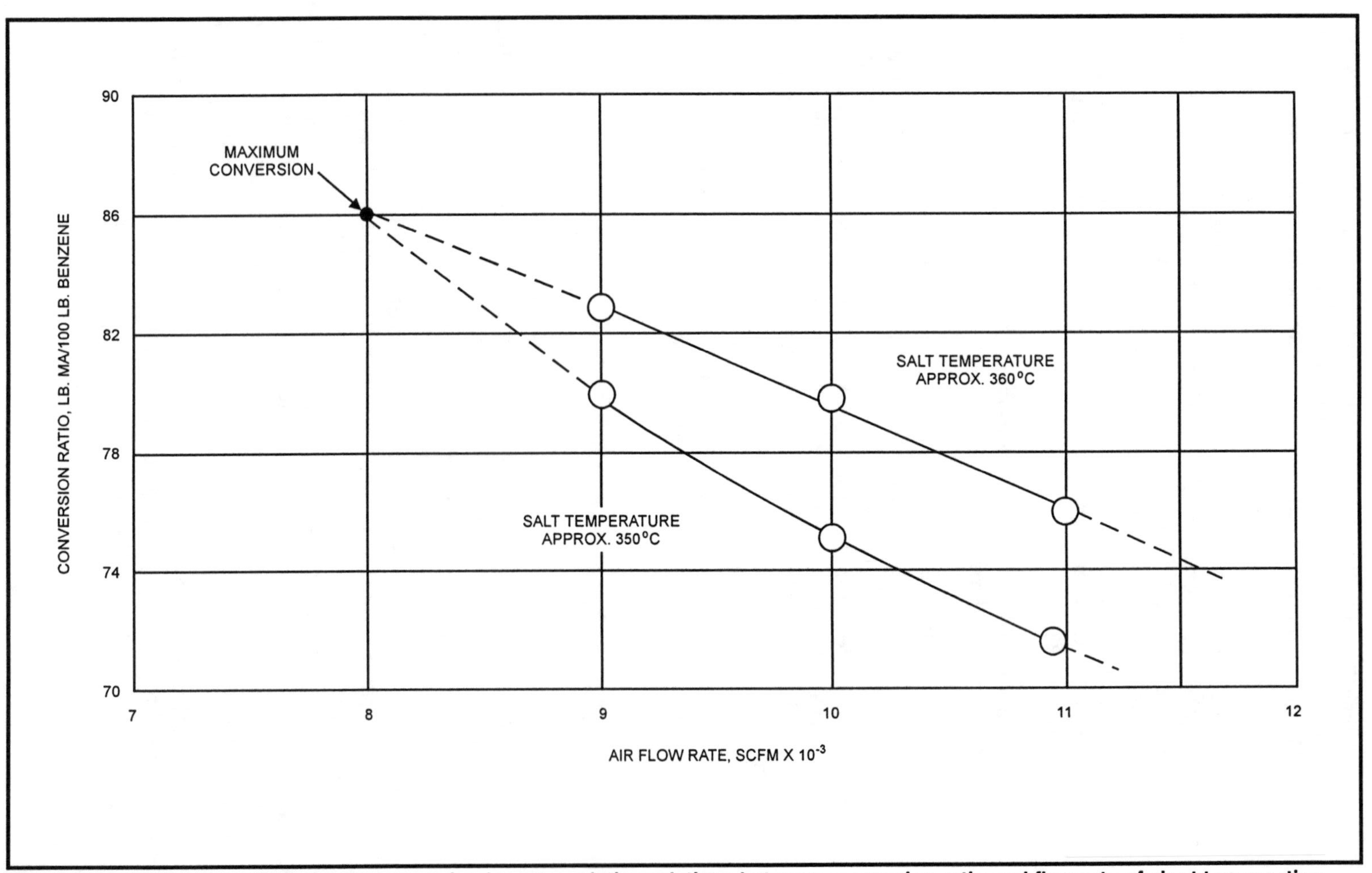

Figure 5-11. Estimation of maximum conversion by extrapolating relations between conversion ratio and flow rate of air at two reaction temperatures

Some Dynamic Response Data

Results of two dynamic tests are presented in Figures 5-12 and 5-13, more as items of curiosity than of utility. This part of the plant was in the manual mode during these tests. Figure 5-12, shows the dynamics of heat absorption in the reactor as the flow rate of heat exchange medium is changed in a pulse-like fashion. Note that the output, the rate of energy removal, does not return to its initial value. The trapezoidal approximation of the Fourier transform routine was used to reduce the pulse data to frequency response form. The frequency response data fit the form K(1 + Ts)/Ts, suggesting a proportional plus integration response. Obviously the input pulse was not brief enough to excite dynamics in a higher frequency range. The apparent integration occurs because, upon increasing the flow rate of coolant, a small but permanent increase in the rate of heat removal from the reactor occurred, accompanied by a slight decrease in temperature of the coolant. This caused a small sustained diminution in the rate of reaction.

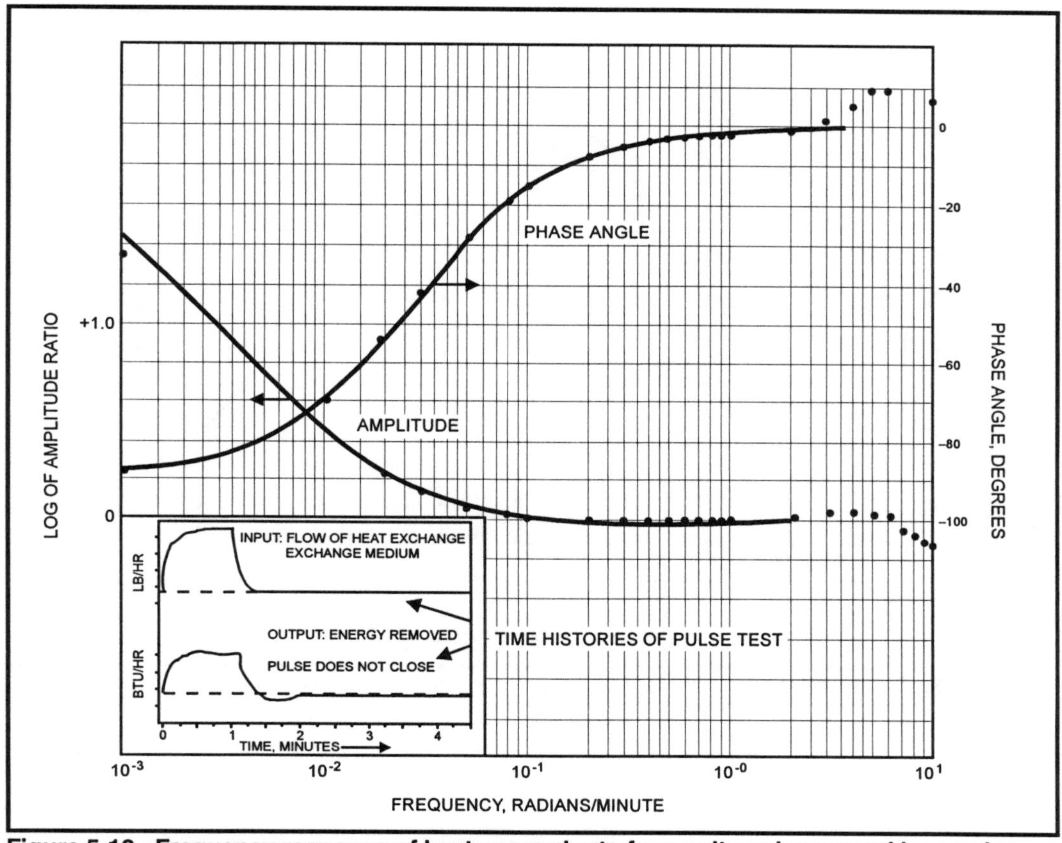

Figure 5-12. Frequency response of heat removal rate from salt cooler caused by a pulse increase in the flow rate of heat exchange medium

Figure 5-13 shows the response of flow rate of liquid maleic anhydride from the after-cooler, caused by a negative pulse change in rate of benzene to the reactor. Again, an integration is present along with both numerator and denominator components, in the linearized approximation of the frequency response information as shown on the Bode plot. The apparent dynamic relationship is considerably more complicated than that shown in Figure 5-12, but the presence of pure integration is also evident. This effect is produced by the brief diminution in flow rate of reactant that results in a diminution in the temperature of the reactor sufficient to decrease the rate of liquid maleic anhydride production, which never recovers. Control systems associated with the reactor were in the manual mode during the procurement of transient response data.

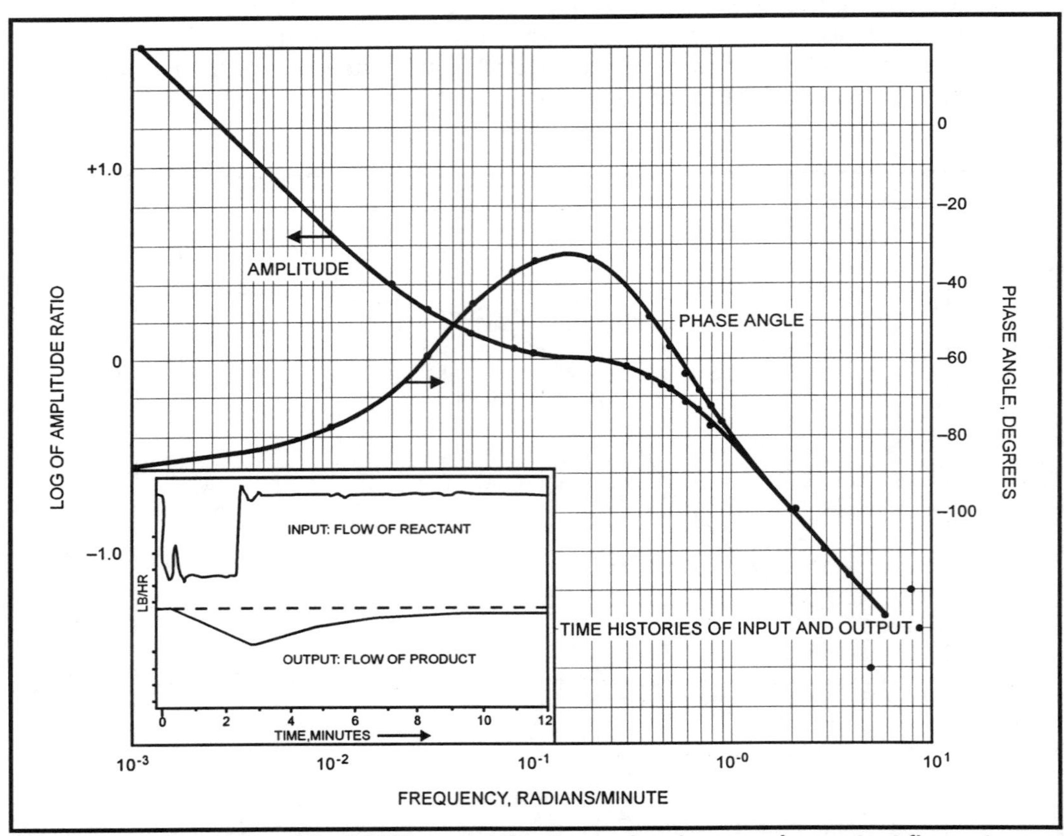

Figure 5-13. Response of liquid MA production to a pulse decrease in reactant flow rate

Summary

Experimental studies of full-scale processing facilities can be carried out safely and with reasonable investments in time and resources. When independent variables are few (say, less than eight) classical procedures can usually lead to problem solution more effectively and efficiently than from statistically designed experimental protocols. For the case illustrated here, instrumentation and procedures for sustaining optimum operating conditions could be easily implemented at minimum cost. Some interesting dynamic responses are shown merely to illustrate that such data can be procured and may, in some cases, yield useful information and insights.

6
Anaerobic Digester for Treatment of Industrial Waste

Many industrial biological processes involve anaerobic fermentations. Synthesis of some chemicals are more economically accomplished by the metabolic reactions carried on by specific organisms than by strictly chemical methods. Processing of municipal waste, for example, is largely dependent upon creating an environment favorable to a wide spectrum of microorganisms that obtain nutrition and energy from this substrate, thereby consuming undesirable pollutants. Other waste material may require rather specific organisms in both aerobic and anaerobic environments. Some industrial wastes can be treated rather effectively by such biological procedures.

The Process

Figure 6-1 shows the salient features of an anaerobic digester and associated gas handling apparatus, which was the subject of this study. The digester is a cylindrical vessel about 85 feet in diameter and 45 feet high. Industrial waste enters near the bottom through a number of channels by which the influent is distributed. Near the 35-foot level, the liquid flows into several collecting troughs leading to large conduits connected to a manifold that conducts the overflow to the surge vessel. The majority of the treated material is recycled through the digester, with about 10% being diverted to aeration basins at another location.

Since the gas generated is suitable as fuel this is directed to the gas manifold serving the entire plant. As seen in Figure 6-1, the gas, after passing through the positive displacement compressor, may flow through the check valve into the gas main, (provided the pressure exceeds about 16 psi). Since it was important to avoid pressure in excess of a few inches of water to develop in the digester, the control system shown in Figure 6-1 had been installed. As will be noted, digester pressure was controlled by regulating the flow rate of gas from the compressor discharge back to the digester through the recycle valve (labeled R, in Figure 6-1).

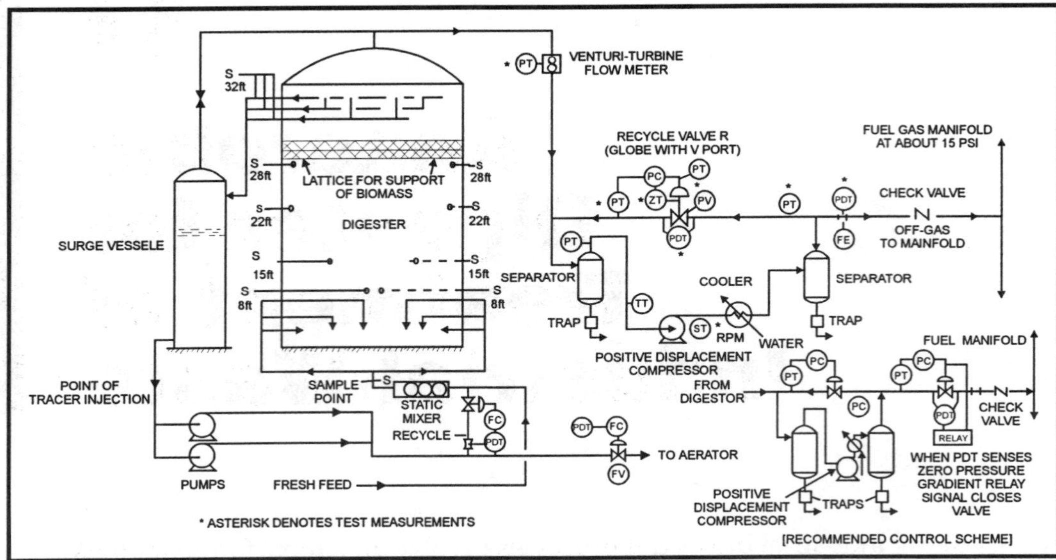

Figure 6-1. Anaerobic digestion system treating industrial water

The Problem

Two areas were of concern:

a. The extent and location of the biomass within the digester. There was some speculation that some of the biomass may have been displaced into the space above the effluent toughs.

b. The poor operation of the gas removal and digester pressure control system, which did not always prevent gas from flowing from the plant fuel manifold back into the digester.

The first was important because the location of the biomass affects the efficiency and efficacy of the digester; the second, because excessive pressure in the digester was hazardous.

The Data System

The data system consisted of two 2-channel Hewlett-Packard Model 17505A recorders provided with high-gain amplifiers and 10-inch charts with adjustable speed. Special circuits for signal suppression were added. By sharing these four channels among several measurements, the desired information was collected, although not as efficiently as might be desired. As usual, all pressures were measured with strain-gage transducers excited with 10 volts, dc. Outputs from these were 5 millivolts full scale per volt of excitation, thus providing sensitivity sufficient to detect exceedingly small changes in pressure signals.

Experimental Studies of Existing Digester Pressure Control System

Initial observations were directed to the arrangement used to regulate the pressure in the gas-containing space of the digester. Abnormal behavior of the existing control components became apparent by merely observing the performance of the recycle valve. The position of this ball valve with a V-port (designated as valve R in Figure 6-1) was seen to oscillate. Upon installing the test instrumentation, and with the controller operating in the proportional mode, the data shown in Table 6-1 were obtained.

Table 6-1.

Controller Gain	Approximate Width of Recorded Excursion	
	Signal to Valve, ma	Valve Stem Displacement,%
8	0.6	5
25	1.2	10
50	3	15

The electrical signals from the controller to the recycle valve (R), as indicated by the millivolt signal across a 5-ohm resistor in series with the input to the input transducer, never remained constant. The amplitude of these signals was observed to depend upon the gain of the controller. As a consequence, the stem of valve R reversed direction of rotation about every 1/2 to 1-1/2 seconds. Upon placing the controller in the manual mode, these oscillations disappeared. This behavior was traced to a faulty item in the central control system. Personnel from the plant instrument department promptly located and replaced the component.

Further testing of this control system showed that, upon changing the set-point (or reference) of the controller from 2 in. to 5 in. WC above atmospheric pressure, the recycle valve (R) opened immediately. However, in the event the valve opened in excess of about 60%, the pressure in the digester increased rapidly. This demonstrated that fuel gas from the plant supply main, at about 15 psig, was flowing back into the digester. When the control valve (R) approached 50% open, the digester pressure would diminish, and control of digester pressure would resume. This back flow of compressor discharge, along with fuel gas, was observed on numerous occasions. However, with the recycle valve operating in the range from 20 to 60% open, control of digester pressure was acceptable. On occasions when the rate of gas generated from the digester was low and the recycle valve opened in excess of 60%, digester pressure control became completely inadequate. (Data confirming these observations are presented later.)

Another objective for studying the digester gas handling system was to develop an alternative method for measuring the rate of gas produced.

The orifice installed in the compressor discharge line was not calibrated, nor was a small turbine-type sensor installed in the large line conducting gas from the digester. Another objective was to use measurements of gas production and rates of pressure change in the digester to estimate free volume therein in an effort to determine if any biomass might have entered the space above the overflow troughs.

Estimating the Free Space in the Digester

The classical method of calibrating gas flow rate measuring devices depends upon precise measurement of the rate of change of inventory of the gas within an enclosure of known volume, at constant temperature and pressure. Traditionally, the gasometer has been used as the primary calibrating device into which the gas enters, having passed through the meter to be calibrated. A gasometer consists of an inverted cylindrical vessel, with an enclosed top, floating on a sealing liquid and carefully counterbalanced so that, as the vessel moves vertically to accommodate the gas entering the open end, constant pressure is maintained. The rate of flow can then be easily computed from the volume accumulated in a measured interval of time.

Alternatively, if the rate of pressure change and flow rate of gas at constant temperature into or from a vessel are known, its volume can be computed.

Two methods were used to estimate the free volume in the digester. One was to compute the volume using dimensions obtained from engineering drawings; the other was to rely on the gas handling specifications of the compressor and measured rates of pressure change in the digester, assuming ambient temperature prevails. The second method is described, and the results are compared to the volume calculated from the dimensions of the digester and associated vapor spaces.

The following relations apply:

- Assuming ideal gas behavior, $pV = nRT$ and, hence, $dn/dt = V/RT(dp/dt)$, if the gas-containing part of the digester volume, V, and the gas temperature, T, remain constant.
- Q_c = volumetric rate of compressor discharge, in cu. ft. per minute, from whence $(dn/dt)_c$, the flow rate in lb. moles per minute, can be computed. The compressor discharge specifications are also assumed valid, and it was operating at 450 RPM, as measured with a tachometer.
- One lb. mole of gas at 492°R and 14.7 psi (407 in. WC) gives R as 297.
- Digester pressure is 3 in. WC or 410 in. WC (abs).
- The flow rate of feed to the digester remains constant as does the rate of gas production from the biomass, during the period of testing.

The procedure is to first close all effluent valves and, with the compressor off, determine the rate of pressure increase in the digester caused by gas production from the biomass. Let this rate of increase be $(dp/dt)_p$. Next open the compressor discharge valve and, with the recycle valve closed, measure the rate of pressure diminution in the digester. Let this be designated as $(dp/dt)_{po}$. Next, find the volumetric rate of gas handled by the compressor. The performance data furnished by the manufacturer (which are given in Table 6-2) can be used to find $(dn/dt)_c$.

$$\text{Then, } V = \frac{RT\,(dn/dt_c)}{[(dp/dt)_p - (dp/dt)_{po}]}.$$

It is the difference in the measured rates of pressure change that determines the volume occupied by gas at the time of observation.

Table 6-2. Gas Handling Capacity of Compressor

rpm	inlet volume, cfm
790	597
585	406
450	278

At the time of the test, the compressor was operating at 450 rpm and, hence, was assumed to be delivering 278 cfm at 14.808 psia and 570°R, or 0.67 lb. moles of gas per minute, thus $(dn/dt)_c = 0.67$.

The time histories in Figure 6-2 were used to find the denominator terms in the above relation. Using the extrapolated lines in this figure, when no gas was withdrawn, the pressure increased at a rate of 6.12/2.10, or 2.91 in. WC, per minute. Thus $(dp/dt)_p = 2.91$. Next the recycle valve (valve R) was closed, and the compressor placed in service, removing gas from the digester. The compressor speed was 450 rpm, as measured with a tachometer. At this speed the rate of digester pressure change, $(dp/dt)_{po}$, was –0.86 in. WC per minute.

The volume of the free space in the digester and associated piping was computed from the relation

$$V = \frac{0.67\,(RT)}{\left(\frac{dp}{dt}\right)_p - \left(\frac{dp}{dt}\right)_{po}} = \frac{0.67\,(297)\,(570)}{2.05} = 55,329 \text{ cu. ft.}$$

This is to be compared to the volume above the draw-off troughs as derived from engineering drawings, 54,067 cu. ft., a difference of 2.8%.

Determination of the rates of gas production from time to time, by the procedure mentioned above, could serve to monitor the biological activity of the digester, as long as the free volume remains constant. The method is simple and requires little in the way of special instrumentation.

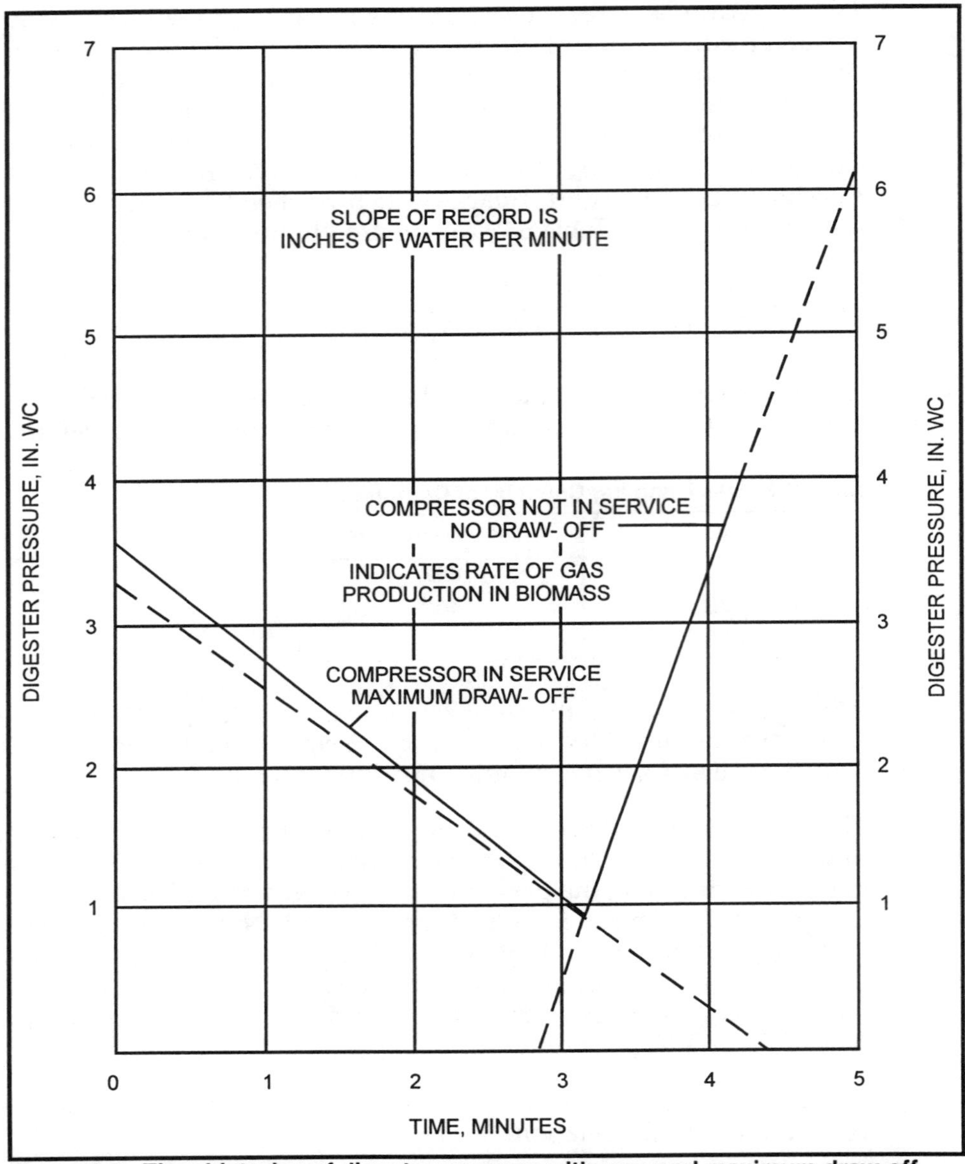

Figure 6-2. Time histories of digester pressure with zero and maximum draw-off

Diagnosis of Recycle Valve Problems

Because of its key role in controlling the pressure on the digester and in directing digester gas to the fuel gas manifold, early in this study attention was directed to the recycle valve (valve R in Figure 6-1). Data shown in Table 6-3 were responsible, in part, for questioning its performance.

Table 6-3. Experimental Data Relative to Recycle Valve Performance

Rate of pressure change in digester, dp/dt in. WC/min	Signal to valve, ma	Valve stem position, % open	Compressor discharge pressure, psi
1.78	4.6	78.4	14.06
1.61	4.8	76.2	9.06
0.544	6.0	61.0	10.46
.531	6.34	57.5	10.46
.345	6.4	57.0	10.36
.118	6.80	53.0	10.96
0	7.00	50.0	10.93
0	6.90	50.0	11.26
–.039	7.2	48.0	11.56
–.333	7.6	43.2	11.71
–.40	8.0	40.0	12.26
–.518	8.1	38.0	13.06
–.567	8.0	39.6	12.16
–1.106	9.54	26.8	14.06
–1.64	max	0.00	15.34

From an inspection of these data, it becomes obvious that, if the recycle valve offers insufficient resistance, the gas from the compressor discharge could flow directly back into the digester.

Another, and more serious, problem apparently existed. The check valve in the line leading from the compressor to the main fuel manifold was not functioning properly. The data in Table 6-3 and Figure 6-3 presents some observations to confirm these conclusions.

Referring to the right-hand side of Figure 6-3, the compressor discharge pressure is about 14 psi when the recycle valve (valve R in Figure 6-1) is 76% open, but drops precipitously to 9 psi as the valve begins to close. Apparently the check valve is not completely effective in preventing back-flow from occurring, as indicated by the high pressure when the valve is open. As will be noted, when the recycle valve is 70% open, gas continues to flow back into the digester, as shown by the positive rate of pressure increase. Of course, gas is being produced in the digester, which will account for some positive change in pressure.

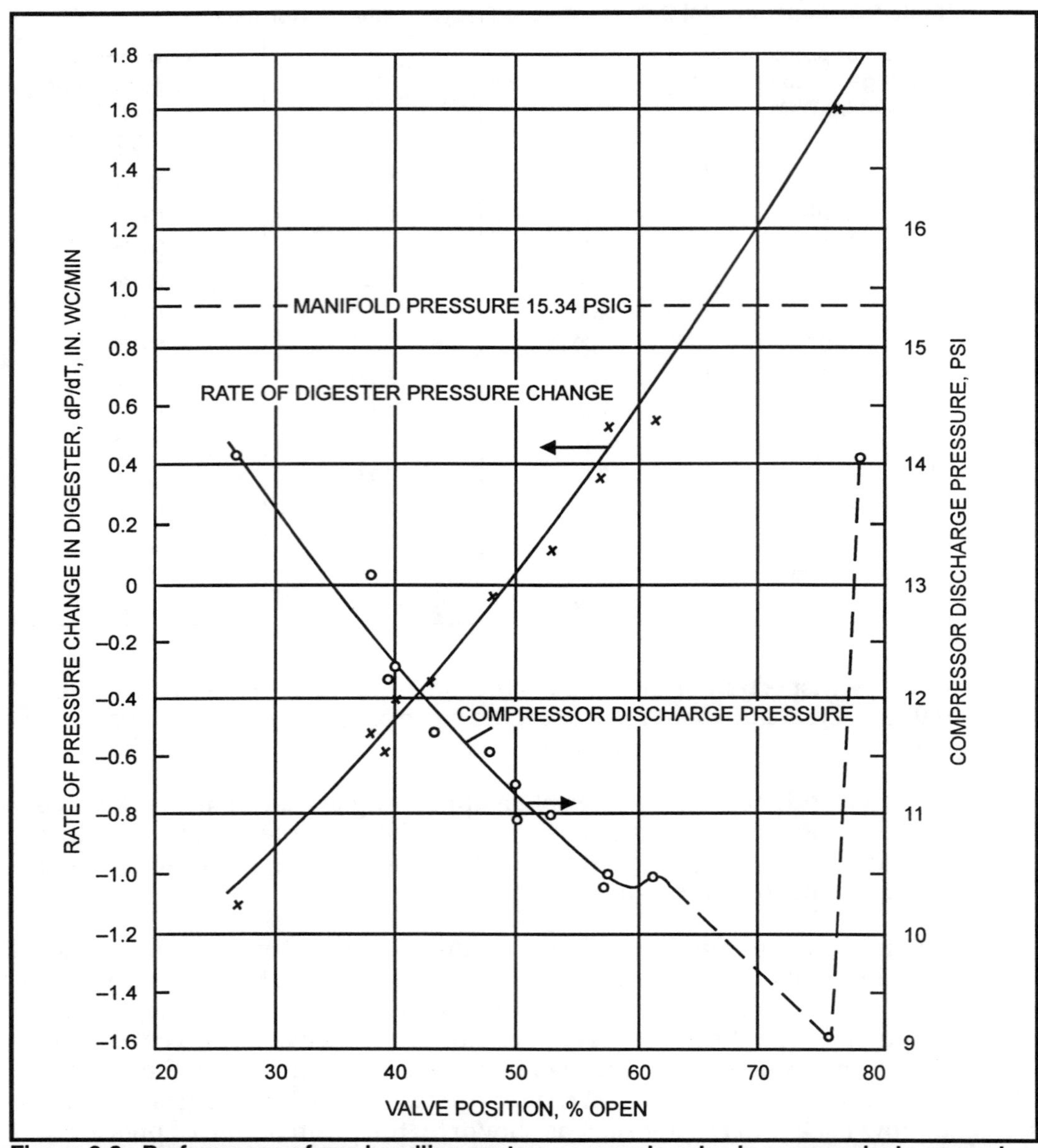

Figure 6-3. Performance of gas handling system as recycle valve is progressively opened

As the valve begins to close, the compressor discharge pressure rises as resistance to flow increases. When the recycle valve closes to less than 50%, the compressor has sufficient capacity to remove gas from the digester and the rate of pressure change becomes negative.

Under the existing circumstances, the useful range of operation of this valve lies between 25 to 60 percent open, corresponding to a signal of 4.8 to 9 ma to the valve-positioning mechanism.

Recommendations for Improving Digester Pressure Control

The uncertainty in the performance of the check valve should be removed by installing a differential pressure controller. This should maintain a pressure difference of about 5 psig, between the fuel gas manifold and the discharge of the compressor. Thus, at all times the compressor discharge pressure will be about 20 psig. The current valve can be used but it should not open more than 50% at any time.

Other Experimental Work

Calibration of Flow Rate Meters

Both the orifice meter used to measure the flow rate into the fuel gas manifold and the turbine meter placed in the center of the large pipe conducting gas from the digester were calibrated using the free space above the liquid in the digester as a "gasometer." The respective relations obtained were:

$$Q = 267\sqrt{(\Delta P_o)/p_a}, \text{ and}$$

$$Q = 21.74mv - 53.9,$$

where

Q	=	cu. ft. per minute at flowing conditions
mv	=	millivolt output from meter
ΔP_o	=	pressure drop across orifice, psi
p_a	=	absolute pressure at orifice plate, psi

Tracer Studies to Determine Extent of Mixing and Free Volume in Digester

The Procedure

For the purpose of estimating the volume of liquid in the digester, a known amount of tracer was injected into the digester. Then, after a sufficient length of time to assure uniformity in concentration and having accounted for all tracer leaving the digester, the volume of liquid in the digester can be computed from a measure of the residual concentration. From samples taken from various positions in the digester during the test period, some insight concerning flow patterns and mixing within the digester can also be obtained.

The Injection Apparatus and Method of Introducing the Tracer

A non-biodegradable material, dissolved in a suitable solvent, was contained in three 55-gallon drums. These drums were placed in a nearly horizontal position, with their outlets connected in parallel to a centrifugal pump. This pump discharged into the suction of the pumps withdrawing material from the surge vessel, as shown in Figure 6-1. Injection was accomplished by quickly opening a gate valve in the injection pump discharge line, and adjusting and maintaining the flow rate at a constant value, as indicated by a rotameter. For this test the time required to inject 321 lbs. of tracer was 13.00 minutes. Note that, because the tracer was introduced into the suction of the process pumps before the point at which material is diverted to the deaeration basin, (see Figure 6-1) a portion of the tracer was eliminated before entering the digester.

Initial Sampling Period to Determine the Quantity of Tracer Injected

Prior to injection and continuing for 20 minutes, samples were taken about every 10 seconds (from the point indicated in Figure 6-1) in order to define the shape of the input pulse of tracer, and to determine the quantity of tracer introduced. (The results are shown in Figure 6-4.) The leading edge of the pulse was very steep, requiring about 1 minute to reach a value of 550 ppm. The trailing edge diminished more slowly, requiring about 2 1/2 minutes. The pulse duration was 16 minutes, compared to 13 minutes for injection.

The area under the time history (shown in Figure 6-4) was calculated by dividing the record into three sections: from 80 to 130, 130 to 890, and 890 to 1040 seconds. The first and last areas were estimated by inspection; the mid-area was computed as (890 – 130) X (mean value of all concentrations in that time span). The result was the equivalent of a rectangular pulse 960 seconds in width and with a height of 425 ppm.

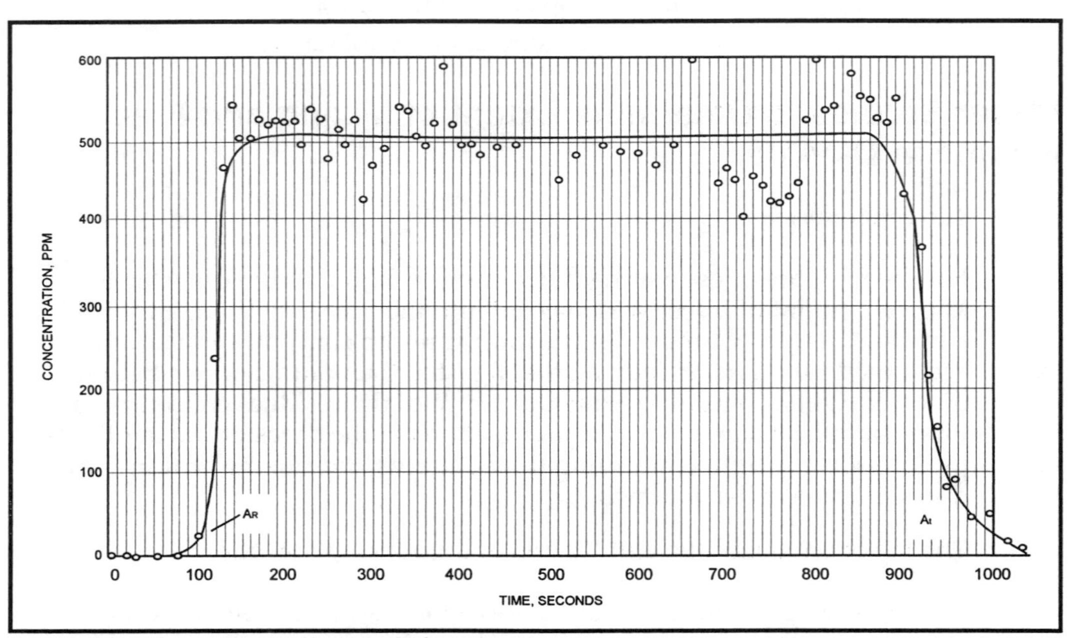

Figure 6-4. Time history of tracer in input pulse

The total tracer entering the digester then becomes

$$\frac{(4950)}{(60)}(8.33)\ (960)\ (425)\ (10)^{-6} = 280 \text{ lb}$$ [1]

This quantity is to be compared to that added from the injection system minus the tracer immediately rejected in the material diverted to the aerobic basins:

$$321 - (321)450/4950 = 292\text{lb., a difference of } 4.3\%.$$

This agreement is considered satisfactory in view of the uncertainty of the analyses and, especially, the industrial flow rate instrumentation.

Evaluating the Consistency of the Data

The consistency of the data may be tested by calculating the recovery of tracer from measures of the concentration in, and quantity of, the effluent. Figure 6- 5 shows the time history of concentration in the recycle (and net effluent) for the first 6 hours. Figure 6-6 extends the time history up to 144 hours, at which time the concentration of tracer has diminished virtually to zero. (Theoretically, the tracer never vanishes.)

Graphically integrating under the record of Figure 6-5 yields 157.9 ppm-hr and from Figure 6-6, 986.4 ppm-hr. The total, 1144 ppm-hr., gives an average concentration of 1144/144 = 7.94 ppm. At a net effluent rate of 450 gpm, a total of $(450)(60)(8.35)(144)(7.94)\times10^{-6}$ = 25.7 lb. of tracer has been removed. This is to be compared to 28.0 lb. as derived from the input pulse data, yielding a disparity of 8.2%.

While not entirely satisfactory, an accounting for 92% of the tracer is considered remarkable and gives credence to the quality of the analytical procedures and sampling techniques used in this study. This also demonstrates that the pulse technique of tracer injection is a viable strategy, useful for studying mixing and concentration patterns in, and even the volume occupied by, liquids in vessels such as the digester under study.

---

1. Flow rate through pumps = 4950 gpm (plant meter). Flow rate to aeration basin= 450 gpm (plant meter).
   Density of liquid= 8.33 lb./gal.

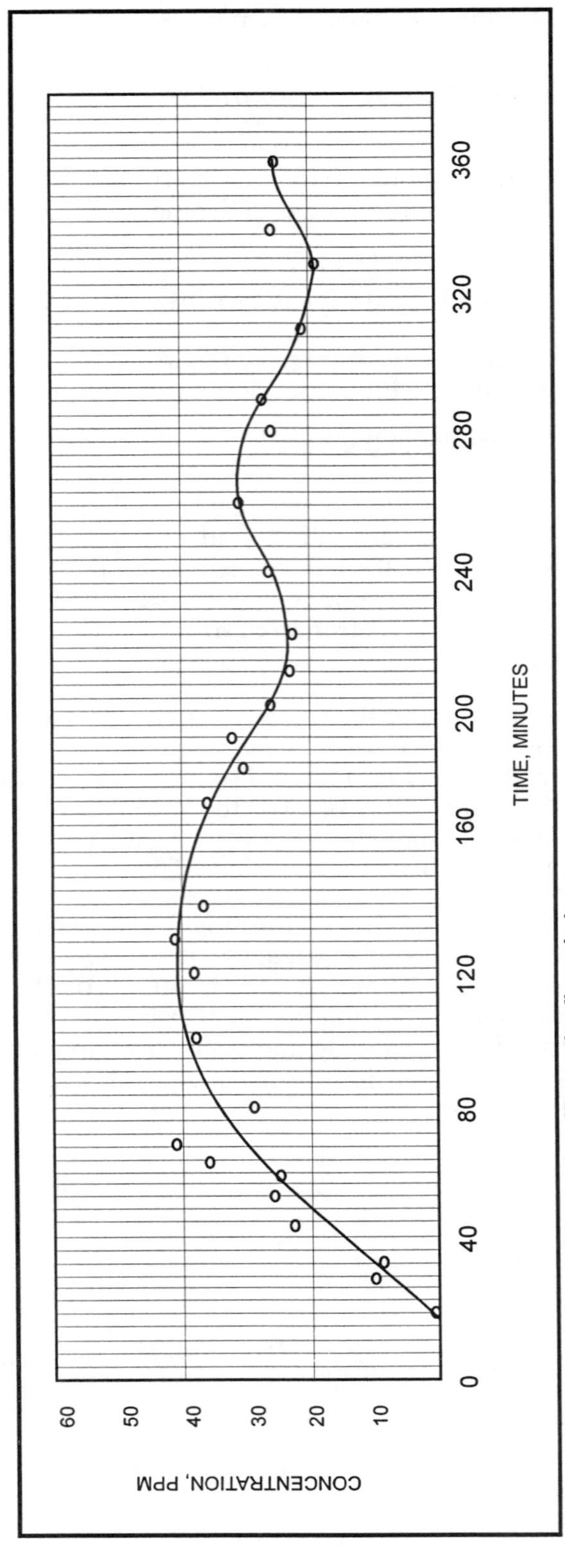

Figure 6-5. Time history of tracer in effluent for first six hours

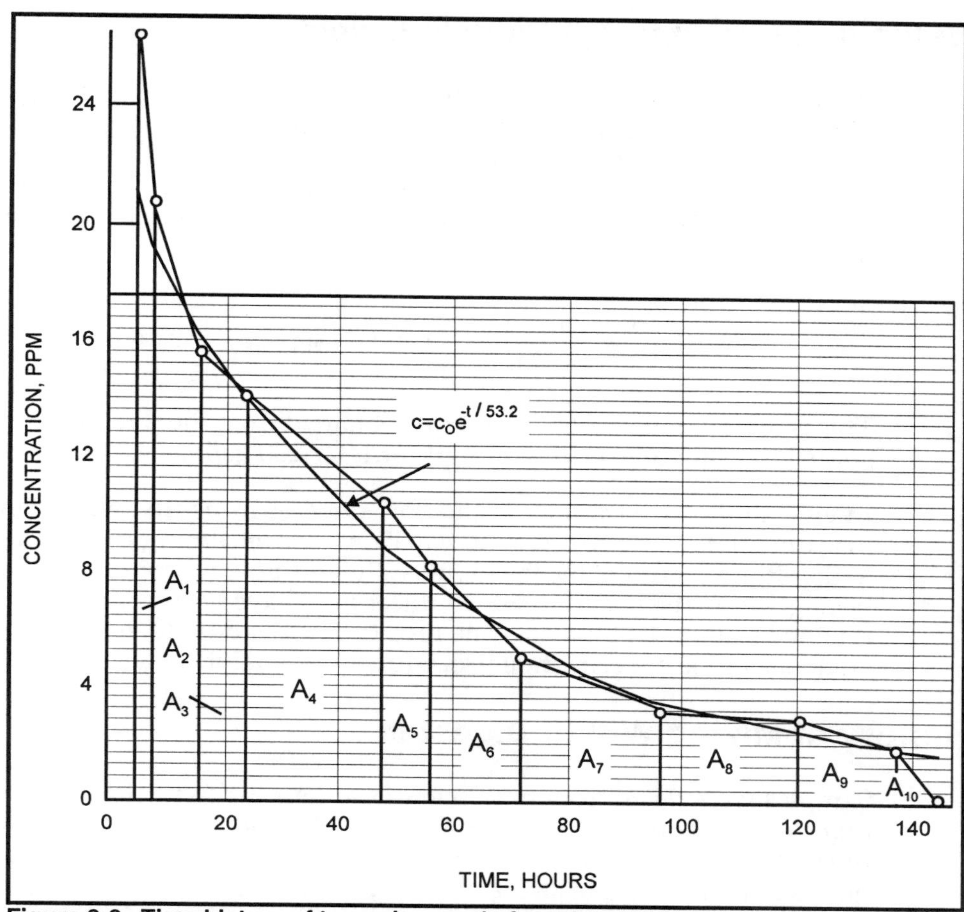

Figure 6-6. Time history of tracer in recycle from 6 to 144 hours

Rate of Tracer Rejection

Data obtained from Figures 6-5 and 6-6 are presented in the semi-logarithmic graph as Figure 6-7. This graph represents the transient response of concentration in the effluent during the first 6 hours following tracer injection. From this figure, an "effectiveness of mixing index" may be derived:

$$\tau = \frac{t_2 - t_1}{\text{in } (c_2/c_1)} = \frac{140}{\text{in } 22/1.6} = \frac{140}{2.62} = 53.4 \text{ hours}$$

The significance of this number is questionable but may serve to characterize the rate at which tracer is removed from a vessel that, in turn, might be related to the intensity of mixing therein.

Estimate of Volume of Liquid in Digester

If the concentration of tracer at the end of 6 hours is taken as 20 ppm (see Figure 6-5), and if uniformity of concentration at that time is assumed, then, if the volume of liquid in the digester is V gallons, the pounds of tracer contained in the digester at that time is $V(20)(8.33)(10^{-6})$. From a previous calculation remaining tracer was 222 lbs. Hence the volume of liquid in the digester is

$$V = \frac{222 \times 10^{6}}{(20)\ (8.33)} = 1,332,500 \text{ gallons}.$$

This is to be compared to the volume obtained from the dimensions of free space beneath the overflow troughs of 1,327,230 gallons.

In view of the uncertainties in the flow rates, as measured with plant instrumentation, no conclusions can be made concerning the location of, and volume occupied by, the biomass. In addition, no consideration has been given to the extent to which tracer may have been absorbed on or retained within the biomass. The results of this study do, however, lend credence to the experimental procedures and analytical techniques.

History of Mixing Within Digester

Samples were taken from two locations at four different elevations in the digester and from the three overflow troughs, for a period of six hours. In the interest of space these time histories are described but not shown. The points at which samples were obtained are shown on Figure 6-1. In all cases the sample probes extended into the vessel about 18 inches and were distributed uniformly around the circumference of the digester.

At the lower levels, up to 15 feet, the tracer showed sharp peaks occurring after 10 to 40 minutes. Thereafter, concentrations declined rapidly. Above 20 feet no peaks appeared; concentrations slowly increased and finally leveled off at about 20 ppm after 6 hours, as did samples from the collection troughs (at about the 31-foot level). The latter showed slightly higher initial concentrations than samples obtained a few feet below, but all approached 20 ppm after 6 hours. Apparently the tracer becomes increasingly dispersed after passing the 14-foot level, and above 20 feet the pattern becomes more uniform.

Perhaps the most important observation is that after 6 hours the concentration of tracer at all sampling points becomes almost identical, in this case 20 ppm. This is the basis for assuming that after this length of time a uniform distribution of tracer has been attained.

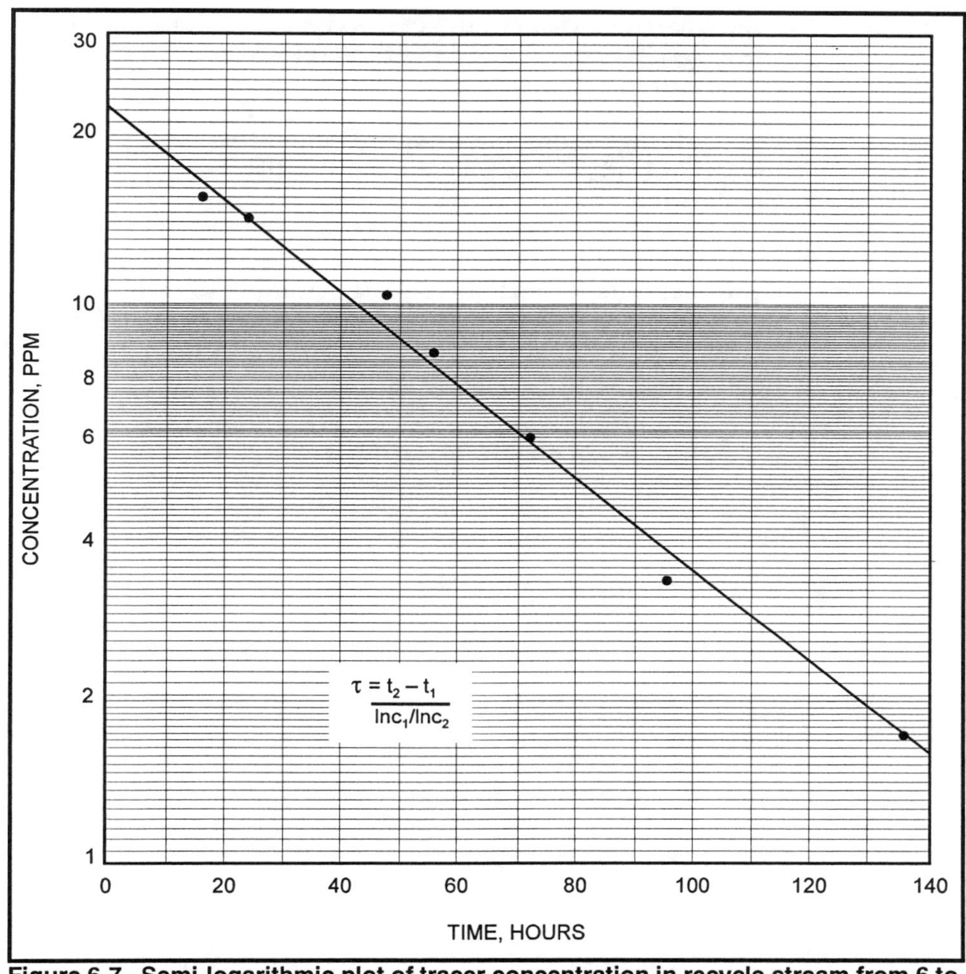

Figure 6-7. Semi-logarithmic plot of tracer concentration in recycle stream from 6 to 136 hours

What Was Learned from the Study

- Check valve performance should always be suspect. To be useful they must operate properly at all times.
- Control valves, especially if in critical services, must be properly sized. This implies that flow rates and operating conditions must be known, along with reliable performance data.
- Orifices and valves in gas-handling service, can be calibrated by measuring the rate of pressure change in vessels of known volume from which the gas flowing through the flow sensor is derived.

- Estimates of void space in vessels can be estimated from rates of pressure change provided gas flow rates into or from the vessel can be measured.
- For studying mixing patterns and estimating liquid volumes in vessels, introducing a tracer in a pulse-like fashion appears satisfactory.
- In any study of mixing phenomena, especially in large industrial equipment, the collection of a large number of samples seems inevitable. The considerable effort in collecting samples and in performing the analyses demands proven techniques, attention to details, and dedicated analysts.

7 The Extraction of a Mineral from an Ore

The Process

Figure 7-1 is a diagram of a process used to extract a metal-bearing compound from an ore. Extraction takes place in two sections of the plant, one at low temperature, the other at a higher temperature. All the processing vessels are very large, hence the inventory of material and energy is also large. The apparatus occupies an area of several acres. This diagram is arranged to emphasize the role of energy circulation and recovery within the system. The items designated as E are tube-and-shell heat exchangers that serve to preheat fresh, highly caustic, extraction liquor. These exchangers receive thermal energy from steam flashed from corresponding vessels V-1 to V-5. Flashing of steam occurs because of progressive reduction of pressure in these vessels. Energy from outside the system enters at two points: with low-pressure steam (at about 110 psi) admitted directly into vessel H-1, and with high-temperature steam (at about 450 psi) admitted directly into vessel H-2. In each case the steam is dispersed through spargers giving rather uniform distribution of the steam within the vessels. Residence time is provided at lower temperatures in reactor R-1 shown at the center, and at higher temperatures in the series of vessels R-2 to R-5 shown along the right-hand side of Figure 7-1. These vessels are about 50 feet high with a diameter of about 8 feet, thus the inventory of material as well as energy in each is considerable.

A slurry of pulverized ore is prepared in the rod mill and pumped by J-1 to heater H-3, which receives flashed steam from V-5. Preheated fresh liquor from the tube-and-shell heat exchangers, E-1 to E-5, is combined with the slurry from heater H-3 and enters heater H-1. Steam from two sources, flash vessel V-5 and the 110 lb. psi source, is added directly to the liquor in this latter vessel. From this heater the liquor enters the low temperature reactor, R-1, at about 300°F. The effluent of this reactor is

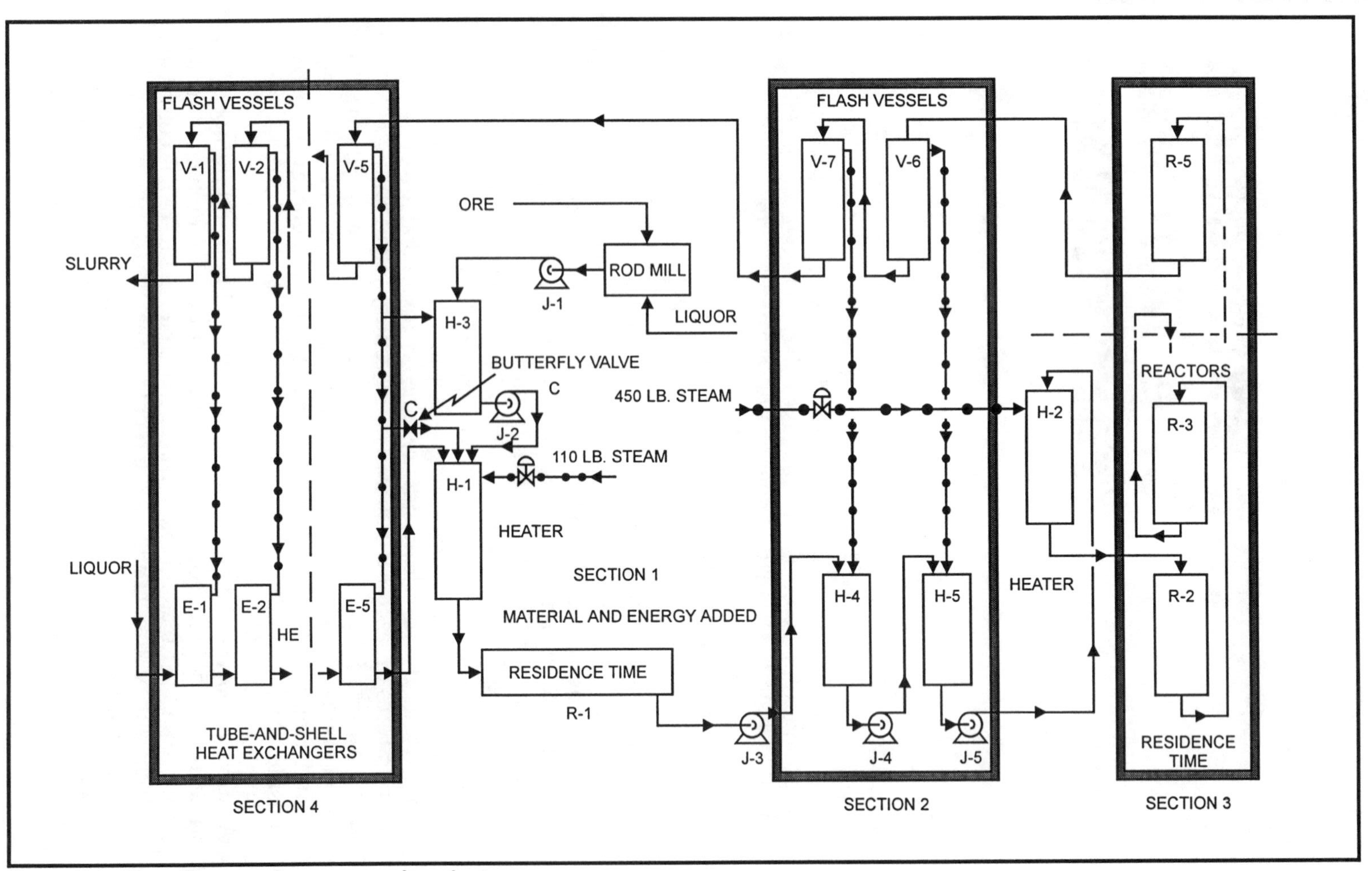

Figure 7-1. Flow diagram of ore processing plant

further heated in contact heaters H-4 and H-5, using steam flashed from vessels V-7 and V-6. The final prescribed temperature is attained by adding 475 psi steam to contact heater H-2; no further energy is added. By progressively reducing the pressure on the enriched liquor, steam is released; this is used, as mentioned previously, to heat the incoming material, thus making the system regenerative in a sense. Inert non-condensable gases created in the process are vented at various points not shown in Figure 7-1. Elimination of non-condensables, especially from tube-and-shell heat exchangers, is most important due to the severe reduction in heat transfer caused by their presence.

The Objectives of the Study

The motivation for considering an experimental study of this process was to improve control of the operations and to reduce processing costs. Major costs associated with the extraction process are the steam used for heating the slurry and the removal of water from the slurry in subsequent operations. Reduction in steam consumption requires that heat recovery be maximized; this can be accomplished by increasing the heat recovered in the tube-and-shell heat exchangers, E-1 to E-5. This implies that steam consumption should be kept to a minimum and that as much steam as possible be removed as condensate from the tube-and-shell heat exchangers. This would reduce not only the cost of steam but also evaporation costs in later processing by which the desired product is produced.

The Problem

A study of this seemingly complicated process showed that the two principal components, insofar as process control was concerned, were the contact heaters H-1 and H-2. It is to these vessels that energy is added from outside the system; all other energy exchange is internal to the process. It follows that if the temperatures of the effluent liquor from H-1 and H-2 are controlled, both the low-temperature and the high-temperature sections of the process can be regulated. Furthermore, by controlling the temperature of liquor from H-1 heater, disturbances arising in the low-temperature section may also be attenuated.

Three sets of pumps, shown as J-3, J-4, and J-5, are particularly important since the capacity of these pumps determines the flow rates of the slurry and, thus, the rate of ore being processed. The flow rates through these pumps were dependent to a considerable extent on their discharge pressures as determined by the downstream processing apparatus. Avoiding large changes in pressure in the vessels into which these pumps discharge was, therefore, important.

The enriched liquor leaves the terminal flash vessel for further processing in other parts of the plant that were not included in this study. The liquor, after being reconstituted, serves as the fresh liquor supplied to this, the extraction process.

This processing complex, with the many paths for the flow of energy and material, presented an imposing picture. The number of variables involved and the numerous ways in which interactions may occur appeared to discourage any hope of discovering useful relationships or to obtain a satisfactory understanding of the operations. Furthermore, to observe cause and effect relationships, using the installed industrial instrumentation, was soon discovered to be quite impossible. Existing instrumentation was either not available or, if present, was neither sensitive nor accurate enough to provide useful information; it was simply impossible to gain any appreciable understanding concerning plant behavior from the existing plant instrumentation. The major difficulties were the inability to see the time relationships between cause and effect and to vary chart speed and sensitivity of the installed recording instruments. Industrial instrumentation is usually installed to serve a specific perceived need, whereas a data system for test purposes must be flexible and able to fulfill very specific demands, not all of which are known initially.

The Data System

Being unable to determine from existing instrumentation which variables were the most significant and to ensure procurement of all needed data, a maximum number of measurements, with a special data system, was necessary. Hence, in the exploratory phase of the test program a total of 55 variables were measured. These measurements included levels in, pressures on, and temperatures of inlets and outlets of all heaters and flash vessels. To accomplish this, with only 24 channels of recording oscillographs, necessitated presenting more than one variable on a single channel by multiplexing. To avoid damaging the galvanometers, it was required that each pen should remain on the chart between samples, *i.e.*, when the input was zero. This severely limited the useful range of amplification. Despite this limitation, some valuable information, mostly of a qualitative nature, was obtained from the extended data system.

A significant result of the exploratory tests is shown in Figure 7-2. These are copies of oscillographs obtained during a test in which the flow rate of high-pressure steam was increased (about 10,000 lb. per hour) for a period of about 6 minutes, with the existing control systems in the manual mode. The temperature of heater H-2 effluent responds rather abruptly, but quickly decreases below the initial value and then oscillates, with diminishing amplitude, for a period of about 32 minutes, eventually approaching the initial value. The record of temperature from reactor R-2 suggests that addition of steam to heater H-2 is felt at its exit some 12 minutes later. No effect can be discerned in the effluent from reactor 3. Whereas the oscillations are becoming progressively attenuated, there is no evidence that they were not present prior to the test.

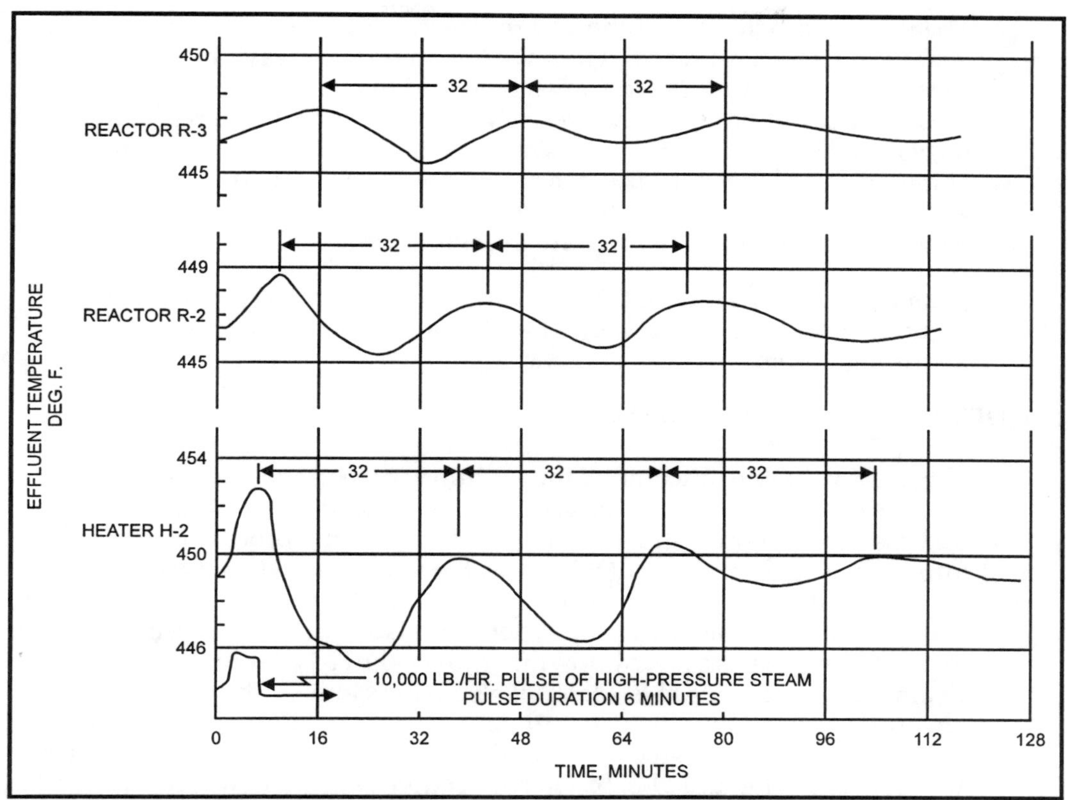

Figure 7-2. Response of heater and reactors to a pulse of high-pressure steam

An explanation of these records, obtained when the controller of steam to heater H-2 was in the manual mode, may be that following the substantial increase in flow rate of H.P steam to heater H-2, the pressure therein increases. This produces a reduction in flow rate through pump J-5, and possibly an increase in flow rate of H-2 effluent to R-2 and to a lesser extent to R-3. This may also account for the somewhat later increase in the temperature of R-3 effluent. Upon returning the steam flow rate to its initial value, the flow rate through pump J-3 increases. Thus, the material, which may have cooled slightly, resumes a higher rate of flow.

The temperature change induced in heater H-2 effluent appears in R-2 effluent about 2 minutes later and in R-3 effluent about 10 minutes later. The sustained, but decaying, transients have equal periods of 32 minutes.

The added energy continues to migrate and circulate around the system, including the low-temperature section. Periodic changes in the pressure drop across the high pressure steam valve must have been created (and there was evidence, from other tests, that transitory increases in temperature oscillations could occur) despite the fact that the valve was restored to it original position.

The preliminary tests also showed that vessel levels and pressures in the vessels, heaters, and reactors followed in an orderly way; that is, as the pressure in the vapor space above the liquid in a given vessel increased, the inventory of liquid in that vessel would decrease. Since regulation of levels was not possible, the amount of useful information that could be obtained from level measurements appeared very limited. As a result, when multiplexing was discontinued, most of these measurements, as well as measurements of pressure in several vessels, were eliminated.

It is important to mention that the variations in temperature shown in Figure 7-2 could not be discerned with the installed industrial instrumentation.

Summary of Initial Observations

The most valuable outcome of the preliminary tests was the insight into process performance that was acquired, leading to the arrangement of the process flow diagram shown as Figure 7-1. Here the process is divided into four sections and arranged to assist in visualizing the flow of material, and especially energy. In the center the raw materials enter, as well as the majority of the extracting liquid. Energy is also added to this section in the form of low-pressure steam flowing into heater H-1 and high-pressure steam entering heater H-2. Material also enters here in the form of fresh ore and its accompanying extracting liquor. Section 2 emphasizes the recycling of energy associated with the steam from flash vessels V-6 and V-7 to vessels H-4 and H-5. Similarly, Section 4 shows how energy is transferred, by way of flashed steam from vessels labeled V-1 to V-5 to the tube-and-shell heat exchangers labeled E-1 to E-5. In Section 3 residence time is provided for the reaction at higher temperature to proceed with energy from high-pressure steam entering this part of the system through heater H-2. Finally, Section 4 (on the left in Figure 7-1) shows how energy is recycled from the flash vessels, V-1 to V-5, preheating the incoming liquid extractant in the tube-and-shell exchangers, E-1 to E-5.

Critical Variables and Their Control

This entire system can be simplified in a "block" diagram as shown in Figure 7-3. In this diagram all streams carry energy as well as material, however "signals" labeled "energy," when associated with this diagram, can be used to enhance understanding of the behavior of this process with the labels indicating the principal function of each stream. This diagram serves to suggest that if the temperatures at points A and B are controlled, the circulation of energy through the various paths, at constant material flow rates, becomes simultaneously regulated. Thus, the successful control of this process depends on eliminating or suppressing the coupling between and among the various parts of the plant by achieving control of temperature at the above-mentioned critical points. If this is done, the result should be a reduction of a complex interacting system into two essentially single-input single-output subsystems for which the design of control systems becomes greatly simplified.

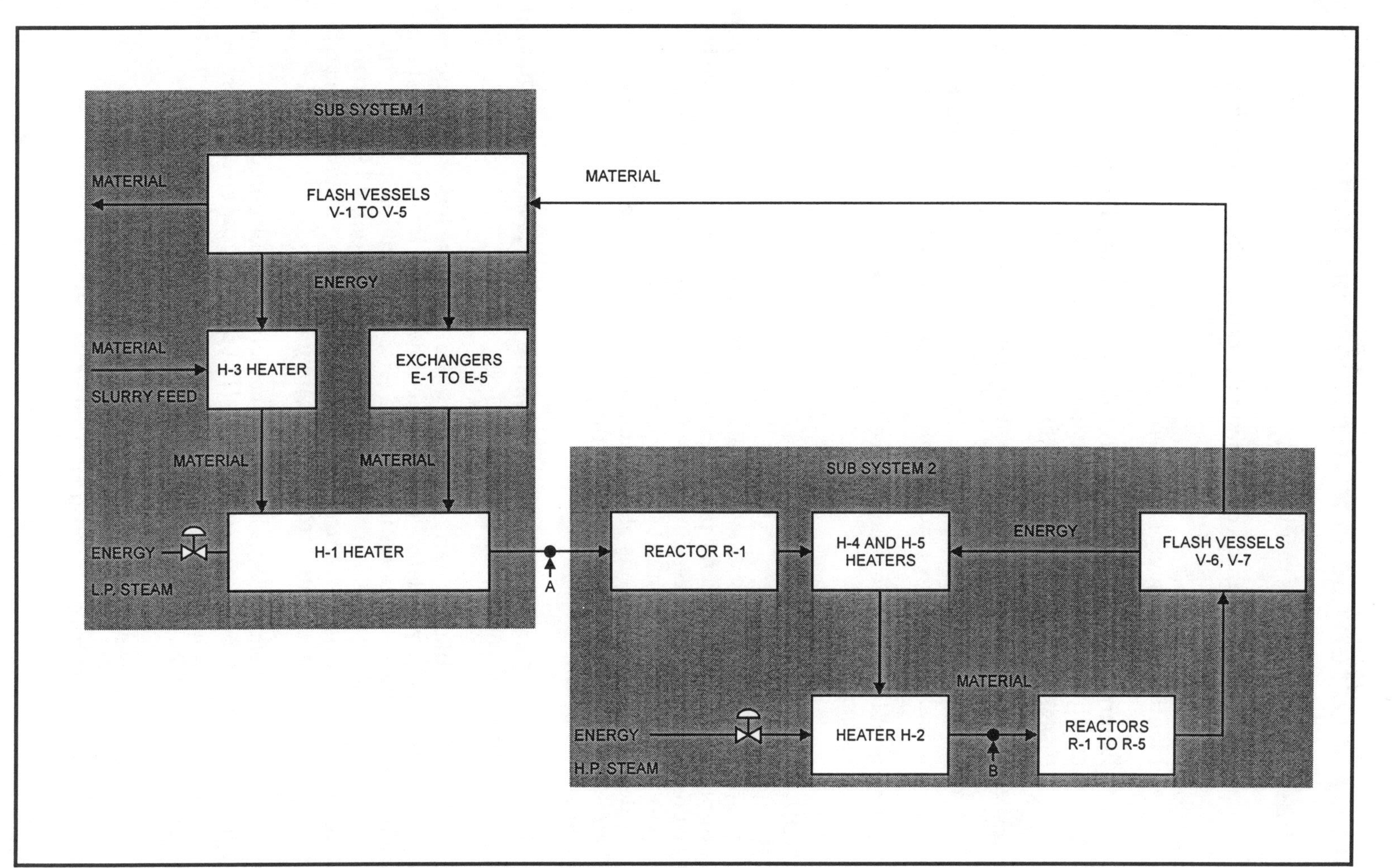

Figure 7-3. Block diagram showing principal material and energy paths

Summary of Observations from Exploratory Tests

These preliminary tests resulted in:

- Developing intuition concerning the physical operation of the process.
- Identifying the most significant variables.
- Determining the quality of signal conditioning (filtering, amplification, suppression) required to yield acceptable measurements from the data system.
- Pointing out particularly sensitive or vulnerable parts of the plant.
- Demonstrating that, despite seeming complexity, the plant was actually relatively simple in its overall behavior.
- Enabled the formulation of a strategy for the control of critical variables.

Dynamic Tests

In addition to those tests in which the transient responses to pulse or step inputs were introduced for purposes of process diagnoses, several tests were made for the express purpose of obtaining specific dynamic performance data. Such data are required for the design of control systems. Discrete portions of the plant were selected, these portions being considered crucial to the control of the process. These specific tests are described below.

Response of H-1 Liquor Effluent Temperature to Change in the Flow Rate of Low-Pressure Steam to H-1 Heater

This part of the plant is particularly important because it is where addition of energy may be regulated, through the flow rate of low-pressure steam. The control of the temperature of liquid effluent from this heater is crucial to the regulation of the plant downstream. Furthermore, if H-1 effluent temperature is closely controlled, effects of disturbances upstream can be attenuated and prevented from propagating ahead.

This test was executed by adding, in an essentially pulse-like fashion, an increment of 110 psi steam to heater H-1. In practice, the steam valve was opened and held in the new position for about 55 seconds, then restored to its initial position. The following variables were of special interest:

- Position of the 110 psi steam valve stem.
- Temperature of the effluent from heater H-1.

- Pressure of heater H-1.
- Level of liquor in heater H-1.
- Pressure drop across low pressure steam flow meter.
- Pressure in V-5.
- Temperature of liquor from heater H-4.
- Temperature of liquor from H-5.

The time histories of two records (the change in the position of the stem of the valve admitting 110 lb. steam to H-1 and the temperature of effluent from H-1) are shown in Figure 7-4. A delay time of 22 seconds has been removed from the output time history so that the output is shown as commencing at the same time as the input. The maximum increase in the flow rate of steam was about 10,000 lbs per hour, as indicated by the plant flow meter. The change in pressure drop across the steam flow meter orifice plate, as measured by a test pressure transducer, was also recorded. No attempt was made to compute the flow rate of steam through this meter because the meter installation did not strictly comply with recommended practice and, furthermore, the reliable metering of steam with a low pressure-drop orifice is fraught with uncertainties. On the other hand, the valve stem displacement could be measured, with a linear variable differential transformer, with great accuracy and precision.

A contributing cause of the above-mentioned delay time (22 seconds) was most likely the temperature sensor, a resistance element enclosed in a 1/4 inch sheath inserted into a heavy industrial thermowell.

Inspection of the temperature response immediately indicates an anomaly: *The undershoot in temperature associated with an apparently conservative system!* Similar behavior was observed in the preliminary tests described above and illustrated in Figure 7-2. In this instance it is possible that an increase in steam to heater H-1 results in a decrease in flashed steam from V-5, thus decreasing the heat transfer in tube-and -shell exchanger E-5. Subsequently the temperature of liquor to heater H-1 decreases as well. This would result in a diminution of temperature of liquor from H-1. Thus, after the temperature increase produced by the added steam appeared in the output of H-1, the effects of the cooler material from the heat exchanger and H-3 became apparent in the effluent of heater H-1. Despite this anomaly, the time histories of Figure 7-4 were processed, via the Fourier transform routine, to yield the frequency response shown as a Bode diagram in Figure 7-5. With the delay time of 22 seconds removed, this relation is, where the gain is taken as unity,

$$\frac{\text{Change in H-1 outlet temp.}}{\begin{array}{c}\text{Change in 110 lb. steam}\\ \text{(Valve stem displacement)}\end{array}} = \frac{1}{\frac{1}{(0.0044)^2}s^2 + \frac{2}{(0.0044)}s + 1}$$

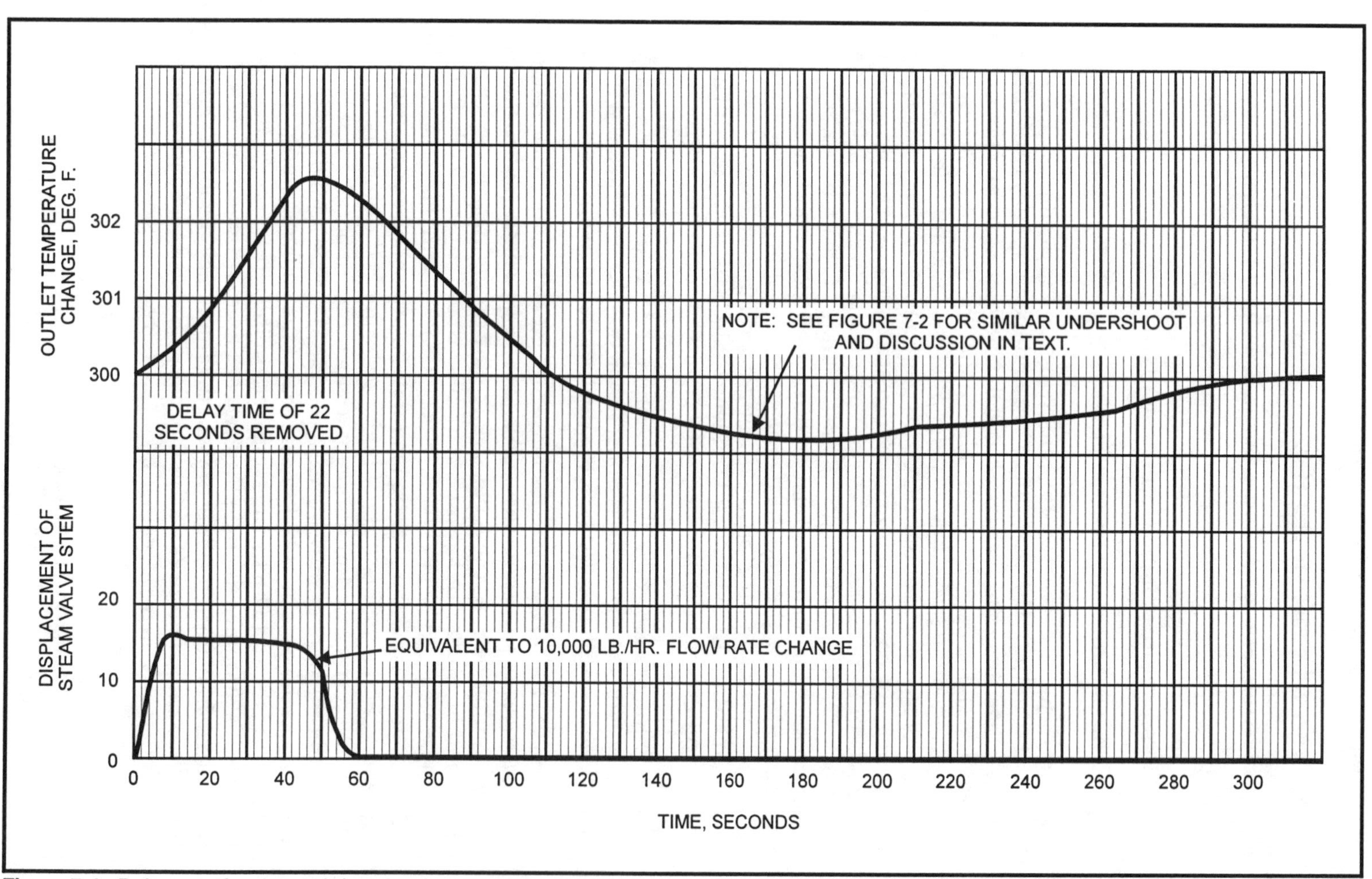

Figure 7-4. Pulse test for determining the dynamic response of heater H-1 to a change in the flow rate of 110 lb. steam

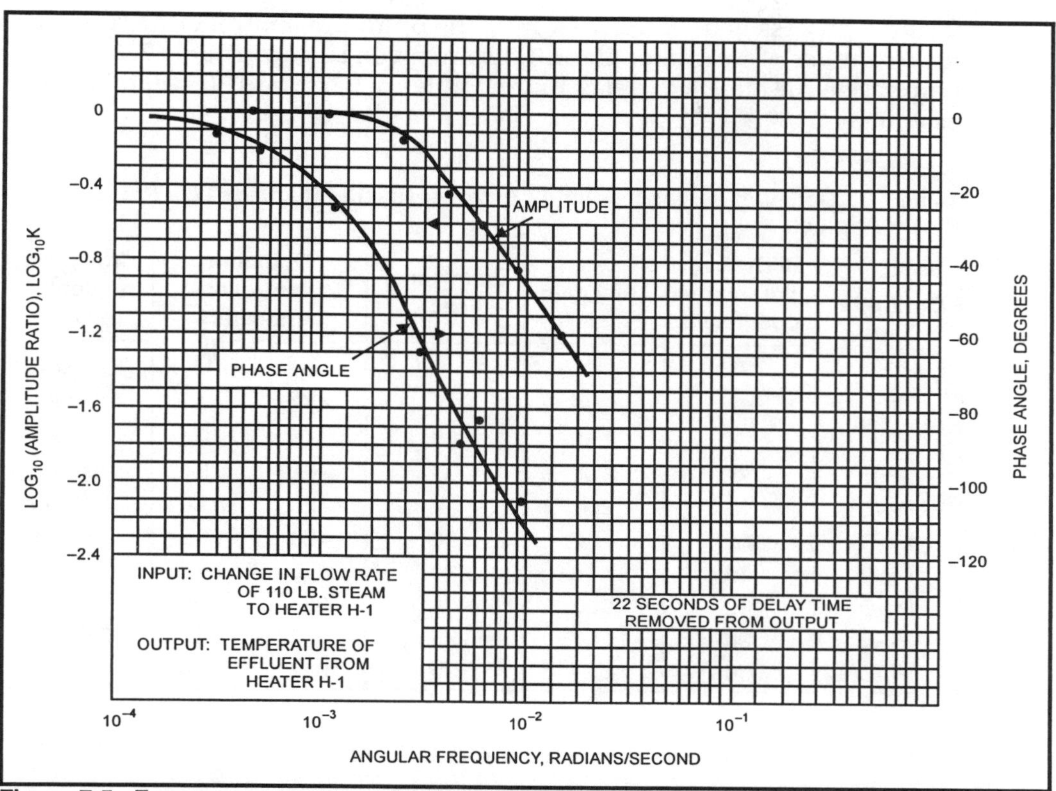

Figure 7-5. Frequency response of temperature from heater H-1 to a change in low-pressure steam flow rate

This is equivalent to a critically damped, second-order linear system having an undamped natural angular frequency of 0.0044 radians (0.0007 cycles) per second.

This relation may also be written as second-order, with two equal time constants, $\dfrac{1}{(1+280s)\,(1+280s)}$.

The pure delay time was 22 seconds; the time constants are in seconds.

Using a process gain of unity (K_p= 1), the optimum controller parameters for a number of controller algorithms are given below in Table 7-1[1]. To emphasize the effects of delay times, values of 0, 20, and 30 seconds are assumed.

Table 7-1. Optimum Parameters for a Controller Used to Regulate Low Pressure Steam Addition

Controller Type	Derivative[1] parameter coefficient	Total gain	Delay time, sec.	Optimum controller parameters		Crossover frequency[2], rad./sec.
	A	K		T_i	T_d	w_{co}
P-I	0	3.00	0	476	0	.0068
P-I	0	2.48	20	479	0	.005
P-I	0	2.09	30	481	0	.0048
P-I-D	10	200	0	426	6.44	.171
P-I-D	10	11.2	20	420	11.6	.031
P-I-D	10	7.46	30	420	13.2	0.23
P-I-D	30	800	0	364	1.96	0.607
P-I-D	30	11.7	20	512	4.34	0.0387
P-I-D	30	7.84	30	512	4.86	0.0286

These numbers were obtained from the software program, PCPARSEL[3].

The efficacy of the several controller algorithms can be judged by the value of the crossover frequency achieved. In particular, note the improvement when a "derivative" component is introduced, as compared to that using only proportional and integration: an increase from 0.005 to 0.031 radians per second in crossover frequency, when the delay time is 20 seconds. Note also that increasing the derivative (see footnote #3) parameter coefficient from 10 to 30 achieved very little increase in the band-pass, when delay time is present. A value of unity for process gain was used because the control system components were found to be uncalibrated and flow meters could not be trusted. When the relations between controller input and controller output, and controller output and valve stem displacement are determined, the correct value of controller gain can be found. In any event, after the correct values of the controller parameters (T_i and T_d) are determined, the user usually finds the best value of controller gain by trial and error.

---

1. The term "derivative" is used advisedly. No pneumatic controller produces pure derivative action, but rather one approximating the operational form, $(1 + ATs)/(1 + Ts)$.
2. This is the frequency at which the closed-loop system crosses unity gain following a peak (of 2 db maximum) for those systems that exhibit this behavior.
3. PCPARSEL is a software program, at one time available from ISA. The curious reader should contact the author

As indicated above, the time histories of the test shown in Figure 7-4 give cause for concern, and hence the results computed from these records, shown in Figure 7-4, should be interpreted with caution. While the records are reliable the interpretation may be misleading. The difficulty is apparent from the temperature record obtained in Figure 7-4 indicating the temperature decreasing about 1°F below the initial value about 200 seconds after the initiation of the test. A reason for this behavior was given above.

Response of Temperature from Heater H-1 to a Change in the Flow Rate of Flashed Steam from Vessel V-5

Since energy to heater H-1 could also be supplied from steam flashed from vessel V-5 through a butterfly valve (at point C in Figure 7-1), the response of the temperature of material from heater H-1 to a change in flow rate through this valve was obtained. This was desired in the event a controller would be installed. To execute this test, the butterfly valve located at point C in Figure 7-1 was opened from 15% to 35% for 25 seconds. The time histories of several variables are shown in Figure 7-6.

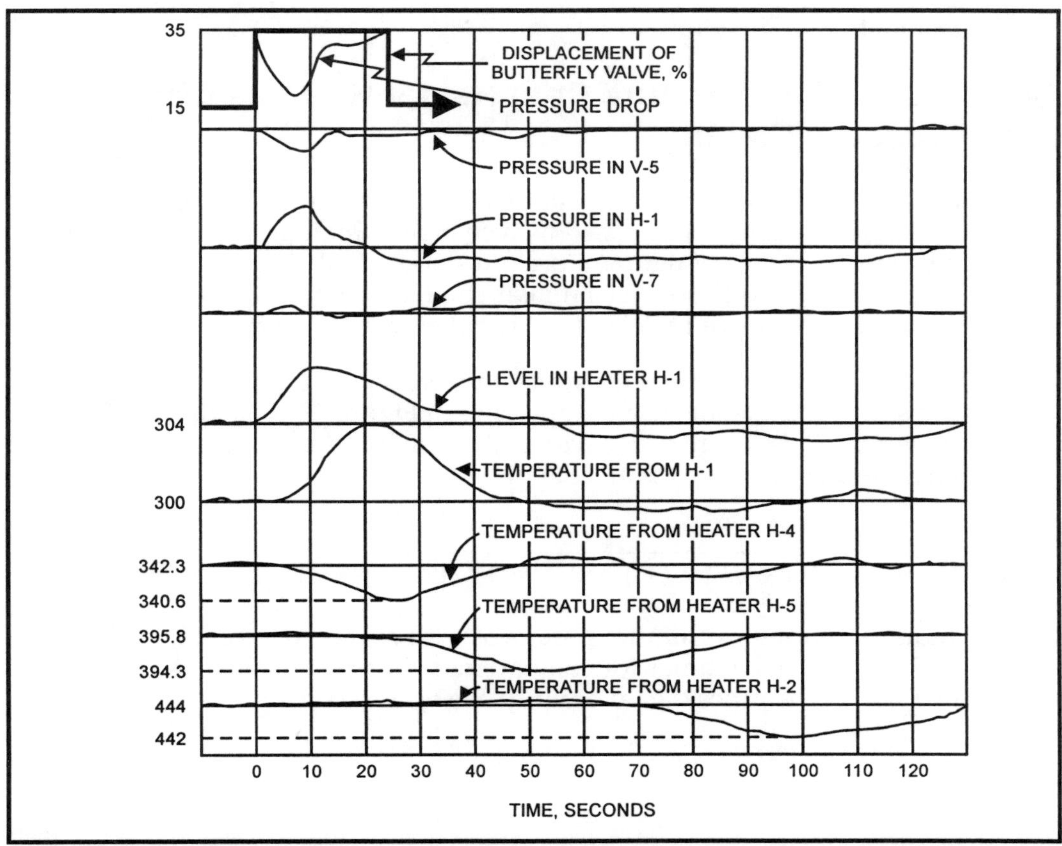

Figure 7-6. Transient response at various points in the process to a change in the flow rate of steam admitted to heater H-1 through the butterfly valve located at point C

While the measured input (valve position) was rectangular, the accompanying change in steam flow rate, as indicated by the pressure drop across the valve, was considerably different. This record is shown in the top left-hand corner of Figure 7-6, where the time history of pressure gradient across the valve appears, together with that of valve stem position. Apparently, as the valve was opened, the pressure difference between V-5 and H-1 decreased, as well as the pressure drop across the valve. As a result, after about 10 seconds, the flow rate of steam through this valve began to decrease sharply. As can be seen, the temperature of H-1 effluent increased in a smooth fashion, showing little or no pure delay time, reaching a maximum excursion of about 3 degrees in about 23 seconds. These changes could not be detected by the plant instrumentation.

Because of the response of pressure drop across the valve (at point C) and inferred diminution in the flow rate of flashed steam as valve V-5 opens, the flow rate of steam from V-5 to H-1 cannot be assumed to increase in proportion to the position of this valve. Using steam from this source may even be counterproductive, and improved performance may result if all the flashed steam from V-5 were directed to the tube-and-shell exchanger E-5. This information may have implications beyond this particular installation in this process.

Response of Temperature from Heater H- 2 to a Change in High Pressure Steam

Figure 7-7 presents the time histories of a test executed to excite the dynamics of heater H-2 caused by a change in high-pressure steam to this device. The record of temperature is a smooth bell-shaped curve responding to an almost rectangular change in the flow rate of 450 lb. steam. The presence of delay time is not evident from these records.

From these data the frequency response was computed using the Fourier transform routine, the result appearing in Figure 7-8. The rapidly falling curves indicate a high order system that can be approximated by the linear form (with unity gain) shown below.

$$\frac{1}{(1 + 530s)\ (1 + 530s)\ (1 + 530s)}$$

Table 7-2 presents optimum parameters for several control algorithms with several delay times for comparison; these results were derived from PCPARSEL. A process gain of unity has been assumed.

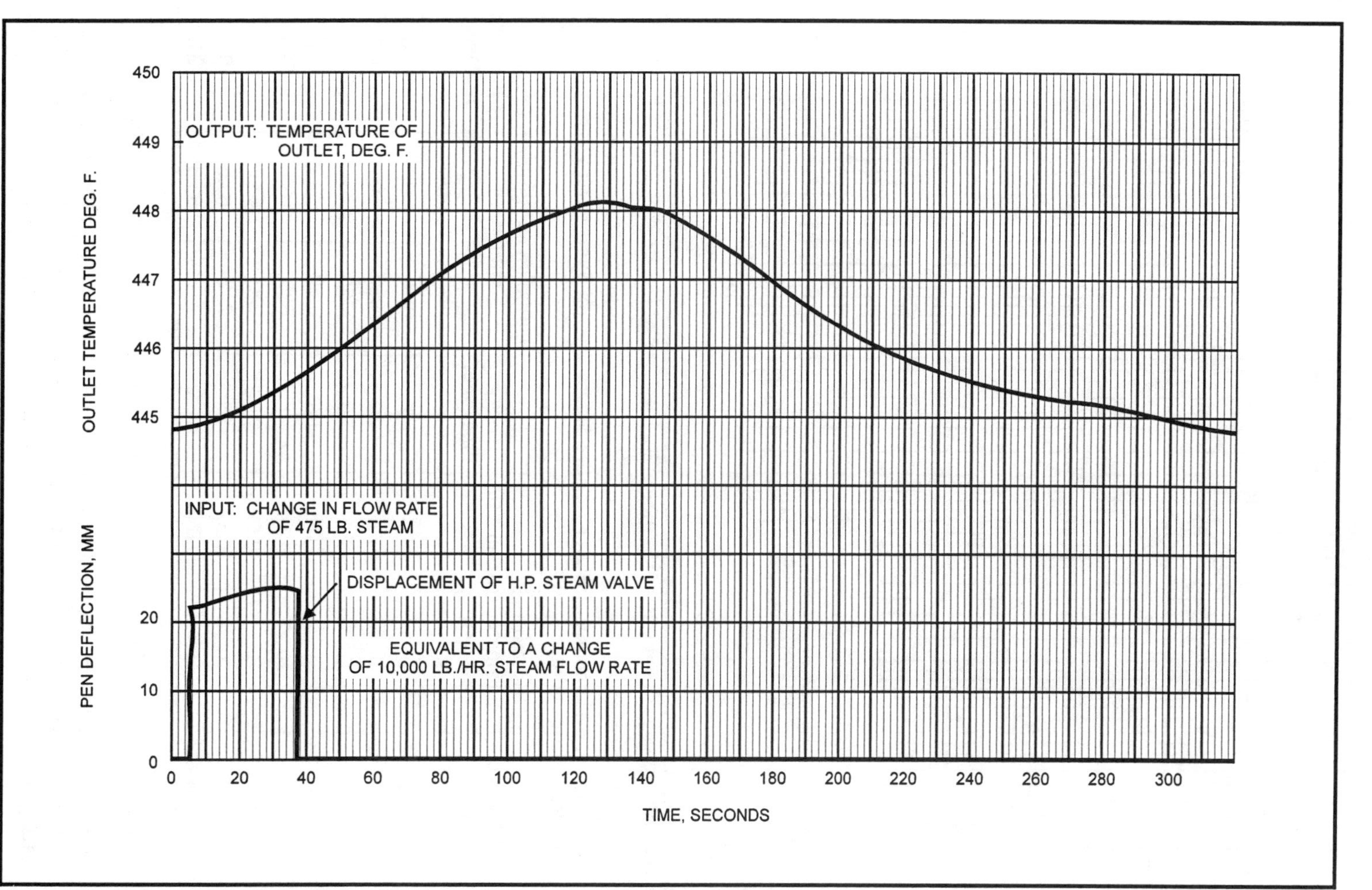

Figure 7-7. Pulse test for determining the dynamic response of effluent from heater H-2 to a change in the flow rate of 475 lb. steam

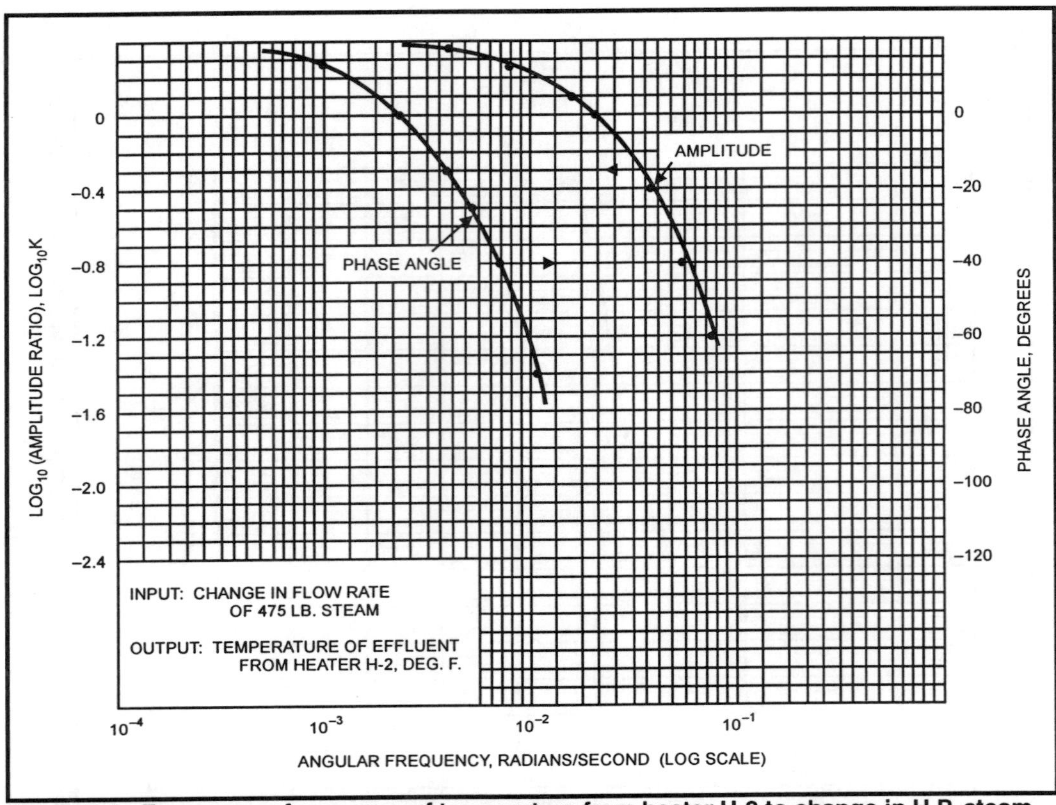

Figure 7-8. Frequency of response of temperature from heater H-2 to change in H.P. steam

Table 7-2. Performance of High Pressure Steam Control System with Various Controller Algorithms

Controller type	Derivative parameter coefficient A	Total gain K	Delay time, sec.	Optimum controller parameters		Crossover frequency, rad/sec. W_{co}
				T_i	T_d	
P-I	0	1.3	0	1060	0	.0018
P-I	0	2.6	10	1350	0	.0034
P-I	0	2.1	20	1430	0	.0027
P-I	0	1.4	120	1650	0	.0017
P-I-D	10	1.14	0	530	63.6	.0028
P-I-D	10	0.52	10	742	63.6	.091
P-I-D	10	0.32	20	742	63.6	0.91
P-I-D	10	0.065	120	742	63.6	0.91

Table 7-2 shows the improvement produced by using a P-I-D algorithm.

The time histories showed little or no delay time. However, to show the deleterious effect produced on the optimum values by the presence of delay time, various values have been assumed. The effect on closed-loop performance is shown by the diminution in crossover frequency that occurs as delay time increases.

No improvement is gained by using a value of the "derivative" parameter coefficient greater than 10, nor does squaring this component produce an enhancement in this instance.

Tests of Control System Components

The next concern was the valve and controller being used at the time to regulate the flow rate of high-pressure steam to heater H-2. Although this was a pneumatic controller, a diminishing number of which are in use today, the recommendation is made that similar studies be conducted during any investigation of this kind. Unless the performance of all components in a control system are verified, one cannot be certain of the final design.

The valve, a nominal 6-inch size with a diaphragm motor and positioner, was first slowly cycled over a portion of its range to determine the presence of hysteresis or "stiction." There was less than 2%. These measurement were made with a Schaevitz linear variable differential transformer attached to the valve stem. Position changes as small as 1/100 in. could be readily detected.

The valve was then stroked from 20% to 90% open to procure the relation between percent displacement and air pressure on the diaphragm. This was found to be

$$\% \text{ open} = 8.75p - 19.3$$

where p is the pressure on the diaphragm in psi.

Next, the gain or sensitivity adjustment (or "proportional band" [PB]) of the controller was calibrated. This was accomplished by displacing the reference index given amounts from the midpoint of the chart at various settings of the PB adjustment. For values of PB from 10% to 50% the sensitivity behaved normally. However, at settings above 50% the relationship deteriorated, being less than indicated by the dial.

To test the integration action of this controller the "reset" adjustment was fully closed, the pen point moved to a prescribed position, the reset adjustment quickly opened, and the response of output air pressure recorded. The response was a typical ramp, as expected, indicating that the controller function was very close to the usual operational form, $\frac{K(1 + T_i s)}{T_i s}$, where p is the output pressure and s is the Laplace operator.

The integrating parameter, T_i, a function of the reset adjustment, is determined as $K \cdot \Delta T/\text{slope}$, where the slope is obtained from a rectilinear graph and ΔT is the time increment. The results are shown in Table 7-3.

Table 7-3. Calibration of Reset Action of Controller

Dial Setting	Integration Time Constant, T_i min.
0.2	0.12
0.5	0.336
1.0	1.05
2.0	1.79
10.0	7.4

The dial settings had no direct relation to the integrating time constant except, fortuitously, at a setting of unity.

These data emphasize the need to calibrate all components if a control system is to be designed.

Summary

As mentioned, the plant was spread over a large area, so that much labor was required to install the test instrumentation. Because of the nature of the material being processed, no means of positively measuring flow rates was possible, thus preventing the opportunity to acquire material and energy balances or reaction kinetic data. Using measures of the electrical power input to the pump motors was suggested as an indication of flow rates. Indeed, this was implemented by using current and voltage transformers, and a Hall effect component associated with the oscillograph. While power changes could be detected, there was no way to correlate these with real changes in flow rate. These measurements, however, turned out to be useful guides to the plant operators. We also learned that using valve stem positions as measures of flow rate can be misleading, especially if rapid changes are made; flow rates may not always keep pace with valve position since the inlet pressure to the valve may decrease.

Despite these shortcomings in experimental technique and absence of meters for measuring the slurry flow rates, we believe the results unequivocally identified the important variables and indicated how they should be measured, what process changes should be made, and where and how control action should be applied. Studies leading to improved economics could not be conducted until adequate data acquisition and control systems were installed on the plant.

In Retrospect

The evidence obtained from the response of the flow rate of vapor from flash vessel V-5 to heater H-1, through the butterfly valve shown in Figure 7-6, may offer insight into the performance of the process existing at the time of this study. Here it was pointed out that the pressure drop across the butterfly valve, located at point C, rapidly diminished after the valve was abruptly opened to a fixed position. (See the record in the upper left-hand corner of Figure 7-6.) Apparently the inventory of steam in V-5 was rather quickly exhausted and, hence, is a variable and certainly unreliable source of energy.

It is safe to conclude that analogous conditions existed between all flash vessels and the recipients of flashed steam. Indeed, the oscillatory variations in temperatures shown in Figure 7-2 may be caused by the variations in flash rates and consequent non-uniform rates of heat transfer to the slurry.

These observations suggest that the overhead vapor from each flash vessel should rather be directed to a tube-and-shell heat exchanger, such as is done in the preheating sections of the plant. The feasibility and economics of such an arrangement deserves consideration. A steam trap is recommended for removing the condensate from each exchanger. Throttling the inlet vapor should be used to regulate the pressure in the exchanger. This is preferred to using the level of condensate to vary the heat exchange surface and, thus, the pressure of condensation. The latter introduces a severe non-linearity since the rate of accumulation of condensate is much slower than its removal through a valve or trap.

Removing water, in the form of condensate, has a direct relation to the reduction in cost of evaporation in later stages of processing. The assumption is made that the slurry can be pumped through such exchangers.

A Scheme for Removing Inert Gases from Condensables

Finally, to remove the inert gases from condensables, the technique shown in Figure 7-9 should be considered. A small, air-cooled, finned tube heat exchanger is attached vertically to each tube-and-shell heat exchanger. A temperature sensor is inserted at the base of a short section of vertical pipe extending above the finned section. The indicated temperature at this point is compared to the set-point of a proportional controller that regulates the flow of inerts through a valve at the top of the assembly. The set-point of this controller should be slightly below the temperature of saturated condensables (steam in the above process) at the existing pressure.

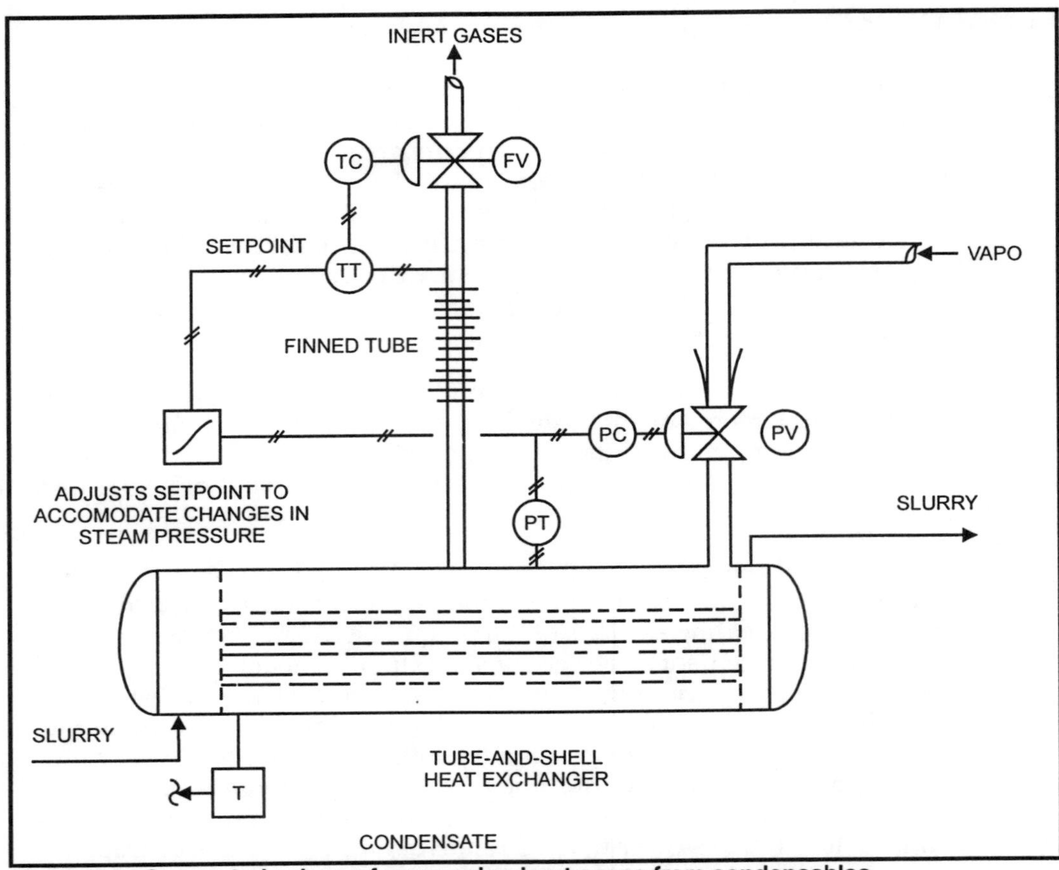

Figure 7-9. Suggested scheme for removing inert gases from condensables

Inert gases, either lighter or heavier than condensables, will accumulate in the space above the finned tubes, and the temperature will decrease as the inventory increases. When the vent valve opens inert gases will be expelled and live steam will reach the sensor, causing the vent valve to close. Further accumulation of inerts will again initiate the cycle.

To compensate for changes in steam pressure within the heat exchanger, a signal from the pressure transducer-transmitter is used to continually and properly change the setpoint, or reference, to the temperature controller regulating the release of inerts.

8

Performance of Boilers and Steam Pressure Control

This chapter presents results of an experimental study of two industrial boilers. A considerable quantity of experimental data was procured; a portion is included to illustrate not only the scope of the study, but also the kinds and quality of information that can be obtained with a flexible data system providing high accuracy and sensitivity.

There is a large body of literature devoted to boiler design, operation, safety, and control. Nonetheless, this study is presented because we know of no similar work illustrating what can be accomplished with a relatively modest investment in time and resources by personnel having no prior knowledge of, or expertise in, boiler technology.

The difficulties in operation prompting this study were described only vaguely, suggesting that control of superheated (S.H.) steam pressure was not entirely satisfactory. Thus, the experimenters were required to collect a large amount of test data in order to ensure that no source of unsatisfactory performance would be overlooked.

Calibration and installation of sensors and preparation of the data system required one week, experimental work another week; the total time spent in the plant being two weeks. Reduction of data, computations, and documentation required several weeks.

The Process

Steam is used not only to supply thermal energy to many parts of a chemical manufacturing complex, but also as a component in some reactions, as well as a source of power for turbines and other prime movers. Boilers are essential features of industrial processing facilities, which must deliver a product of desired quality, under varying conditions to a wide variety of users.

The boilers, three in all, discharged superheated steam into an 18-inch manifold about 1500 feet in length, from which many widely distributed users were served. Figure 8-1 is a process flow diagram of one of the three boilers showing the essentials of the existing control arrangement. It is important to mention that this diagram does not show the interlock system by which the boiler is shut down in the event of a hazardous condition, nor does it show the cross-linking between fuel and air flow rates.The principal purpose of the diagram is to indicate the points at which measurements were made.

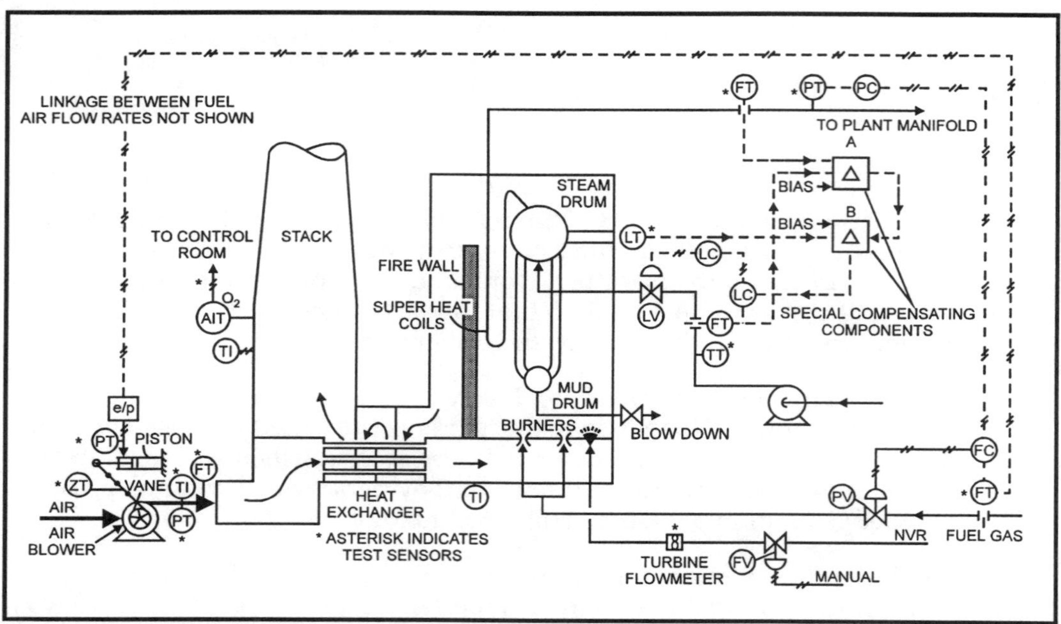

Figure 8-1. Boiler with original controls

The process is not complicated since, except for burning fuels, no chemical reactions occur. On the other hand, mechanisms by which energy from combustion of fuel is transferred to the water are not simple, consisting of a combination of radiation, convection, and conduction. Natural gas comprised the major portion of the fuel; however, a small but variable amount of chemical waste (termed non-volatile residue, or NVR) was also introduced from time to time.

Superheated steam generated from the boilers was supplied to a large number of users. These included direct consumers of 550 psi steam, desuperheaters, pressure-reducing systems, turbines, and heaters distributed throughout the plant complex. The steam distribution system constituted a large volume into which the boilers discharged, this volume influenced the rate at which pressure could change. Figure 8-1 shows one steam-generating system isolated from the remaining portion of the plant.

The control systems in use at the time for regulation of flow rates of air, make-up water, and superheated steam pressure are shown. The method by which the air flow rate is linked to the fuel rate is not shown. The boiler feed water control scheme included two components used to make compensatory adjustments to the setpoint of the feed water control system. This scheme was intended to ensure that if the drum water expanded, as a result of diminution in drum pressure following an increase in steam demand, the water rate would increase, not decrease. Component A measures the difference between steam generated and water entering the drum. If this is positive, the output signal is added to the signal from the boiler make-up flow meter in Component B, causing an increase in feed water. Appropriate bias voltages are inserted so that signals are at appropriate magnitudes. During the test work these components were not functioning; no significant outputs were detected at any time.

Preparing for the Testing Program

Steam generating systems, because of the hazards involved, are customarily provided with "interlocks," which terminate or curtail their operation in response to conditions considered unsafe or any number of other reasons. Because of these interlocks, installation of some sensors, independent of those already in service, was not permitted.

For example, replacing orifice plates with flow nozzles or Venturi tubes to measure fuel and superheated steam flow rates was impossible without plant shutdown. For fear of activating the interlock system, it was impossible to install calibrated differential pressure transducers with which to measure the pressure drop across the installed orifice plates in fuel gas and make-up water lines.

Test transducers were calibrated at the plant site; these included pressure and pressure differential strain-gage-type sensors and the turbine meter measuring the flow rate of NVR. The existing sensor for measuring the flow rate of air was not calibrated, nor were the sensors of miscellaneous temperatures. The accuracy and precision of data derived from plant measuring systems was not ascertained. (The locations from which measurements were procured are shown in Figure 8-1.)

Measurements and Data System

Signals from all test sensors were conducted to recording oscillographs through shielded and well-grounded cable. Simple RC filters at the input of each high-gain, high-impedance dc amplifier enabled the rejection of noise, or signal corruption, at selected frequencies. Two consoles of recording oscillographs of 8-channels each provided simultaneous records of 16 variables. All pressure transducers could follow signals faithfully at 60 Hz and the amplifiers at 120 Hz. Strain gage transducers were excited with 10 vdc, which produced full scale outputs of about 5 mv per volt of excitation. (Sensors and their location are identified in Figure 8-1.)

Testing Protocol

The usual practice of testing operating plants is to first obtain data typical of existing conditions and then to proceed with specific tests on various components and subsystems of interest. The initial tests serve to establish a status quo and, in addition, frequently uncover sources of difficulties not previously suspected. At the least, these tests assist in understanding the processing scheme and strengthening intuition concerning plant behavior.

These preliminary tests should be conducted with as many existing control systems in the manual mode as is feasible. Thus, the behavior of controllers are removed from the scene of action, eliminating the need to know their behavior at this point in the study. If not removed from the closed-loop, the performance characteristics, especially dynamic, of each of the control components involved must be known in order to determine the behavior of the primary processes. Furthermore, inferior characteristics of control components may be found to be major contributors to poor overall performance, so that test information from closed-loop systems would be of marginal value in providing specific information about the essential processing apparatus.

In any event, in this study, after records typical of existing operations were obtained, special tests were conducted on each boiler at various levels of steam production. The inputs used in a given series of tests were changes in the flow rates of fuel gas, NVR, and air to the burners. These changes in independent variables or inputs were of two kinds: positive and negative steps, and positive and negative pulses.

Step inputs are very abrupt changes in independent variables from one level to another. Few process inputs can or should be changed instantly for a variety of reasons. For example, it is neither wise nor feasible to change the position of a large valve instantaneously. Thus, inputs designated as "steps" were in reality more like an exponentially decaying increase or decrease, depending on the direction of the change. Step changes are useful for obtaining the sensitivity of a process or component, this being the relation between output and input in the steady state. In addition, responses to such changes are sometimes useful for estimating major time constants, and, frequently, sources of process malfunctions can identified from the responses.

In conducting a pulse test the independent variable or input is changed in a fairly smooth manner from an initial value to another, then returned in a similar manner to the initial value. Usually dependent variables respond in like manner, although the responses will not always return to their initial states.

From pulse tests definitive information descriptive of dynamic, or transient, performance is obtained. A single properly executed pulse excites the system with all significant frequencies, and the response can be extracted from the data by appropriate computations. In this study, and others described herein, this computation uses the trapezoidal approximation of the Fourier transform. This data reduction technique is described in Appendix B of this book.

Important Measurements

Principal outputs or responses measured during each test in this study were:

- Pressure of superheated steam in the main near the boiler.
- Flow rate of superheated steam from the boiler.
- Temperature of superheated steam.
- Excess oxygen in flue gas.
- Level in steam drum.
- Flow rate of make-up water to steam drum.
- Outputs of several transmitters or plant transducers.

Closed-Loop Tests

To execute these tests other boilers, as well as users of steam, were placed on manual control to eliminate as much as possible any change in steam production from, or consumption by, these components. The objective was to determine how the boiler under test would respond to disturbances with its existing control systems in service. The disturbance in each test was a deliberate change in the rate of fuel to one of the other boilers. These changes in fuel rate were so small that compensating changes in air supply rates were not necessary, excess air always being sufficient. The superheated steam pressure controller in service on the boiler under test was pneumatic with the proportional band set at 100% and the reset adjustment at 5 minutes. No derivative action was present. The mathematical form of the controller algorithm was not verified nor was the instrument calibrated.

Figure 8-2 shows time histories of the response of No. 3 boiler, with its control systems in service, following a reduction of about 1800 lb/hr in the rate of flow of fuel gas to No. 2 boiler over a period of 2 minutes. The upper records in Figure 8-2 show the change in fuel rate to No. 2 boiler and the response of fuel rate to No. 3 boiler. The lower three records, on expanded ordinates, show the response in steam production, which reflects the fuel rate and the pressure of superheated steam.

The response of flow rate of fuel to boiler No. 3, the boiler under test, showed very little change for about 1 minute, although the pressure of the superheated steam was rapidly decreasing. The fuel rate to No. 3 then began to increase for a period of 2 minutes, reached a plateau at 3 minutes, before commencing a slow rise. In the meantime the inventory in the plant steam header, or manifold, has been reduced, so that after 10 minutes, header pressure has fallen from about 550 to 548 psi (lower record). Full recovery of steam pressure obviously did not occur for considerably longer. This behavior was hardly discernible on the plant recorder; the changes in steam pressure, for example, producing a pen deflection of perhaps 1/8 inch on the control room recorder!

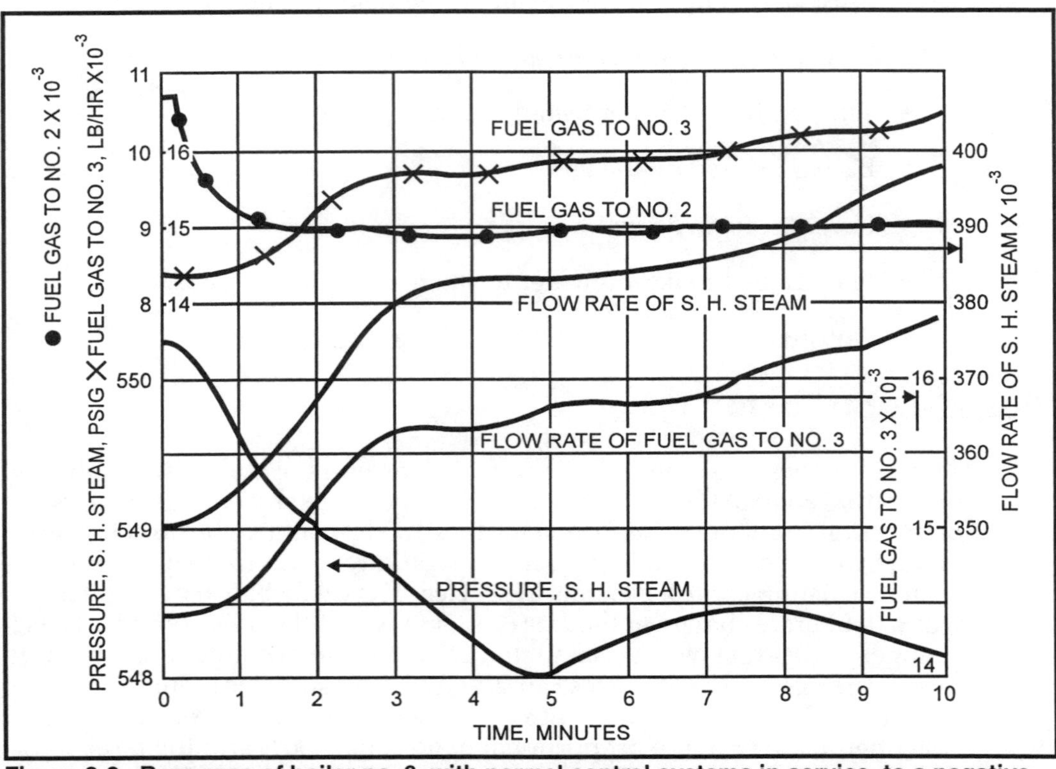

Figure 8-2. Response of boiler no. 3, with normal control systems in service, to a negative step change in fuel to no. 2 boiler

Figure 8-3 presents similar records of the response of No. 3 boiler to a substantial <u>increase</u> in fuel rate, about 4700 lb/hr., to No. 2 boiler. This abrupt increase in fuel did not exceed the capacity of the burners. As can be seen, the flow rate of both fuel gas to and steam from the test boiler (No.3) decreased. Nonetheless, the steam manifold pressure increased about 6 psi before beginning to decrease after 6 minutes have elapsed. Considerable undershoot occurred after 10 minutes.

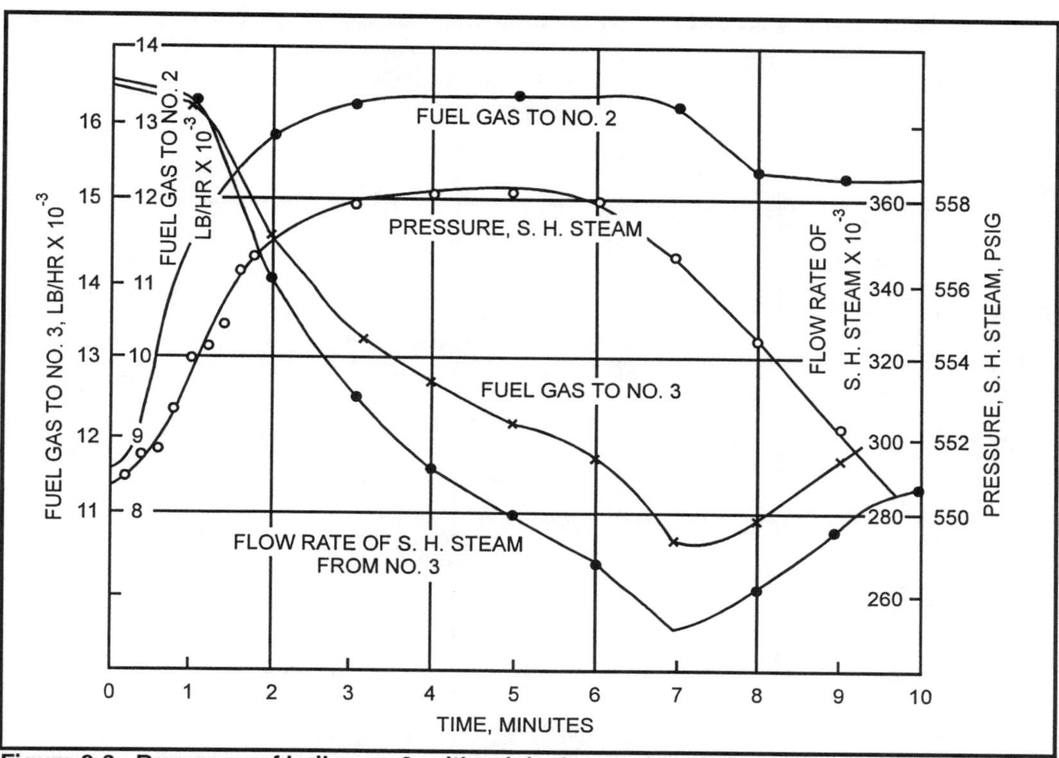

Figure 8-3. Response of boiler no. 3, with original control system in service, to a positive step change in fuel gas to no. 2 boiler

The response of the boiler under test to a negative pulse in fuel to No. 2 boiler is shown in Figure 8-4, where the pulse duration was about 2 minutes. Although the test boiler responds more favorably than in the previous tests, about 2 minutes were required before the steam pressure recovered, following disappearance of the input pulse.

These preliminary exploratory tests, and others not described here, demonstrated that the steam pressure control system associated with No. 3 boiler was not particularly effective, despite the fact that changes in fuel rate can drive steam pressure quite rapidly. Tests on No. 2 boiler revealed similar behavior. Obviously the steam pressure control systems were incapable of producing an appropriate response to either disturbances or setpoint changes.

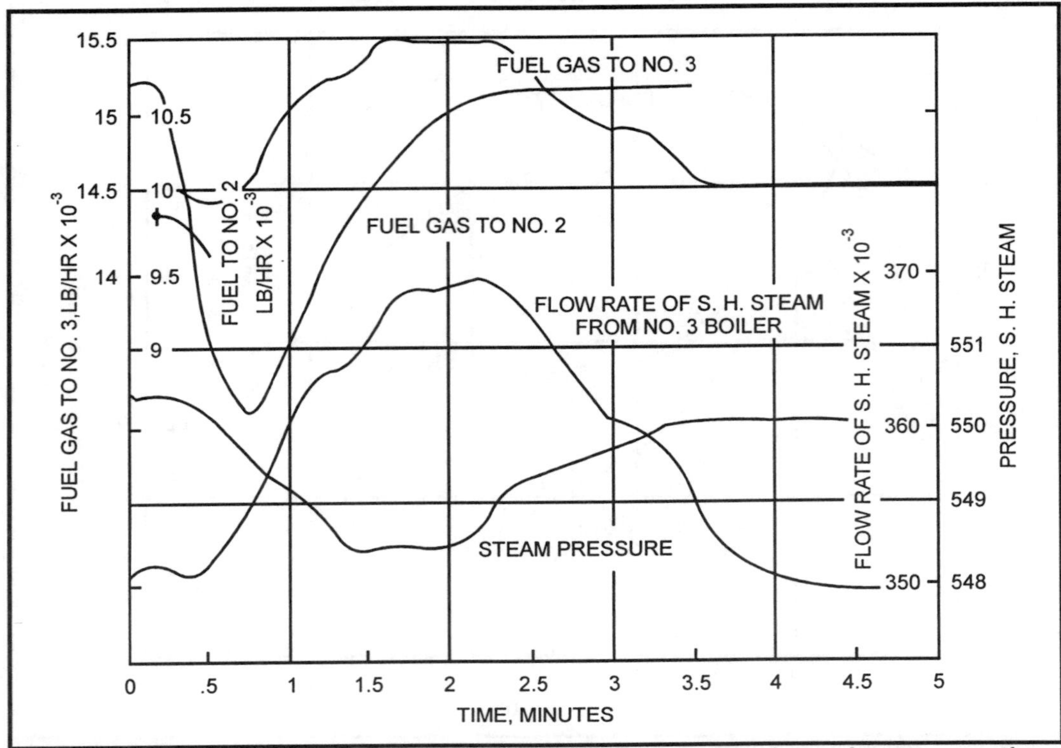

Figure 8-4. Response of boiler no. 3, with original control systems in service, to a negative pulse change in fuel gas to no. 2 boiler

Open-Loop Steady State Tests

Considered here are results of open-loop steady state tests of boiler No. 2. With the exception of steam drum level control, all control systems were placed in the manual mode. These data are particularly useful in demonstrating consistency between steam and fuel rates, as well as giving the value of certain performance indices, such as the ratios of flow rates of steam to fuel gas and air to fuel gas rates. Apparent discrepancies between steam and feed water flow rates are also evident.

Table 8-1 summarizes the results of tests on No. 2 boiler, each set of data presenting the average of several observations at very nearly the same conditions. NVR was included in the fuel at two fuel gas firing rates. The differences in excess oxygen should also be noted.

These results show that, in the absence of NVR, the steam to fuel gas ratio is about 24, and the ratio of air to fuel used was between 170 and 183 scf/lb. of steam. When NVR is used both ratios increase, as expected.

Table 8-1. Summary of Steady State Performance of No. 2 Boiler

Fuel gas, lb/hr	11,121	9,201	9,195	15,200	14,664	13,545	13,545
NVR, gpm	none	8.02	12.92	none	none	7.52	12.10
Feed water, lb/hr	282,739	277,656	280,675	313,786	306,336	313,774	325,590
Air flow rate, scf/hr × 10^{-6}	1.892	2.325	2.235	2.700	2.685	2.700	2.700
S.H. steam, lb/hr	273,389	249,379	256,200	361,591	354,020	358,975	375,590
Steam/fuel gas, lb/lb	24.56	27.10	27.86	23.80	24.14	26.50	27.73
Steam/fd. water, lb/lb	0.967	0.90	0.91	1.15	1.13	1.14	1.15
Air/fuel gas, scf/lb	170	252	252	178	183	199	199

The inconsistency in the steam to feed water ratio clearly points to a measurement-instrumentation problem since, as steam production increases beyond about 300,000 lb./hr, indicated feed water flow rates become less than steam production, an obvious impossibility. Recall that calibrated sensors, independent of plant sensors, could not be installed for measuring all variables of interest.

The relation between steam production and fuel consumption for boiler No. 2 is shown in Figure 8-5. One remarkable feature of the straight line relations in this figure is that the upper two passed precisely through the origin, and the lower line required only minor adjustment to do likewise. Each point on these graphs represents the average of from 2 to 9 steady-state tests. The above relationships between steam and fuel rates can be expressed by the simple relation

$$W = (0.33N + 24.3)F$$

where

W = steam flow rate, lb./hr
N = flow rate of NVR, gpm
F = flow rate of fuel gas, lb.

A significant interesting observation forthcoming from this relationship is the apparent appreciation in the heating value of fuel gas as the proportion of NVR is increased. Thus,

$$\frac{W}{F} = 0.33N + 24.3$$

indicating that at a given fuel gas rate the ratio of steam to fuel gas increases in a linear fashion with the NVR added. Also note that

$$\frac{dW}{dF} = 0.33N$$

signifying that the contribution of NVR increases with NVR flow rate.

A possible explanation for this enhancement, aside from the heat of combustion contributed by NVR, may lie in changes that occur in flame emissivity as the proportion of NVR in the fuel is increased. It should be pointed out, however, that at low firing rates, during which no NVR was used (see Table 8-1), oxygen in the flue gas was between 2.33 and 3.05% (average 2.74%), while at the high firing rates the oxygen content in the flue gas was between 0.34 and 1.06% (average 0.9%).

The effect of excess air was obtained from test data where air flow rate was changed from 2.706 to 3.106 × 10^6 scf/hr, causing an increase in oxygen in the flue gas from 1.03 to 1.75% and a decrease in steam production of 3612 lb./hr, or about 1%. This is equivalent to 1.5% decrease in steam production for each percent increase in flue gas oxygen. Of course, part of this decrease is caused by the increase in energy absorbed by the air. On an equal percent excess air basis, the slopes of the lines in Figure 8-5 tend to become less, although the divergence of the lines still exists.

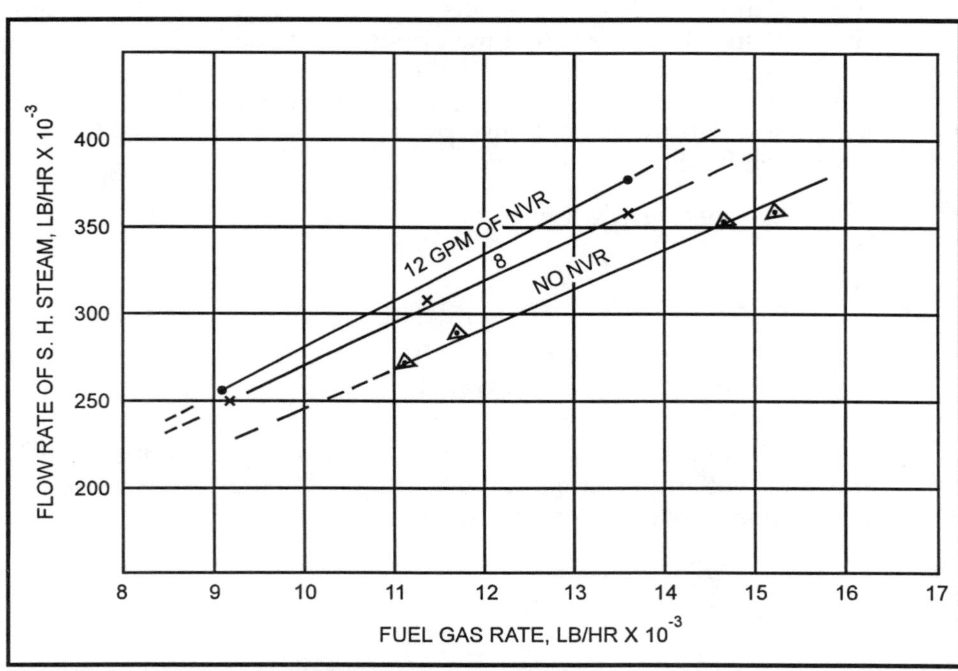

Figure 8-5. Steady state performance of no. 2 boiler

Some fragmentary data on the performance of the air blowers were also obtained from steady-state tests. This information appears in Figure 8-6. The relation between air flow rate and blower discharge pressure was approximately linear over the range for which data were procured. No correction was made for the actual density of air, flow rates being calculated directly from plant instruments using given meter factors. The turbine speeds during these tests were about 3500 and 4900 rpm for boiler No. 2 and No. 3, respectively.

Figure 8-6 also shows the relation between the adjustable throttling blower vane position and air flow rate, over the range studied. These relations are

$$Q = 0.046 \times 10^6 \cdot d \text{ and}$$

$$Q = 0.100 \times 10^6 \cdot d$$

for No. 2 and No. 3 blower, respectively, where d is the displacement of the vane.

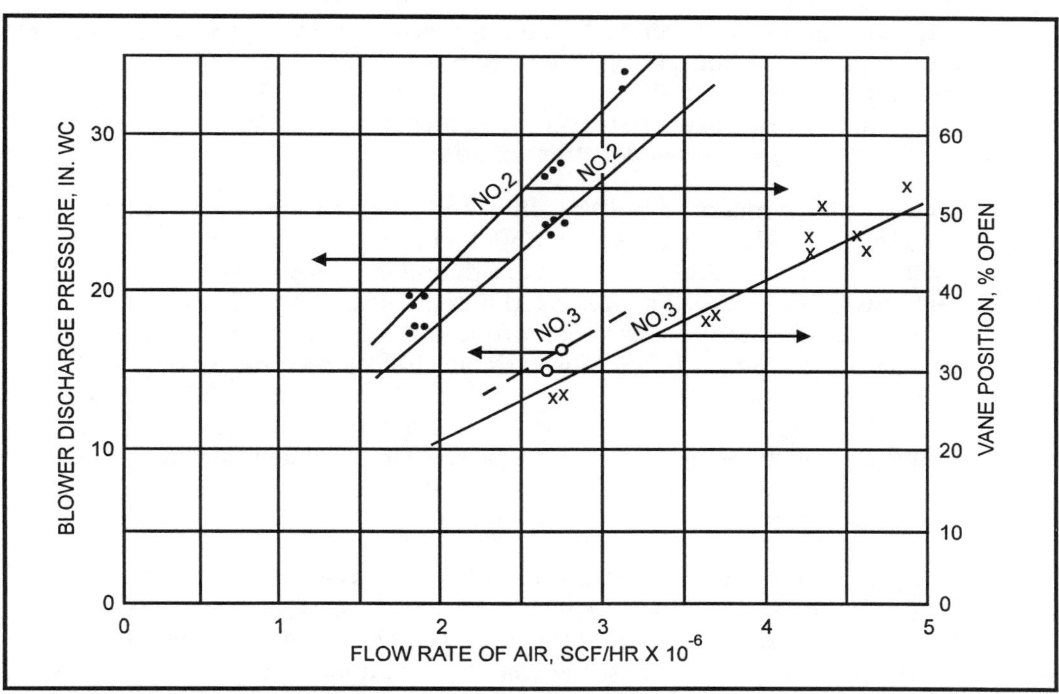

Figure 8-6. Performance of air blowers

Dynamic Tests

The purpose of dynamic tests was to determine transient characteristics. For the boilers under study the inputs (or independent variables) and outputs (dependent variables) considered are given in Table 8-2. A maximum of 24 input-output pairs is possible; not all are important.

Table 8-2. Inputs and Outputs Considered in the Tests of the Boilers

Inputs	Outputs
Fuel gas flow rate	Pressure of S.H. steam
NVR flow rate	Flow rate of S.H. steam
Flow rate of air	Temperature of S.H. steam
Position of air vane	% Oxygen in flue gas
	Air blower discharge pressure
	Flow rate of air to burners

Two types of inputs, as previously described, were chosen to excite the system of concern: steps and pulses. In connection with pulse inputs, it is important to emphasize that responses to such inputs need not close, *i.e.*, return to their initial values. Indeed, if the system being tested possesses the properties of integration, the response cannot return to its original value. Nonetheless, the computer program used to reduce such responses to frequency response must be able to process these data and retrieve valid results. The program used in these studies had this capability.

Dynamic tests included both positive and negative steps and pulses, with tests being conducted at two or more levels of steam production on both boilers. Since transient behavior of both boilers were similar, only test data relevant to boiler No. 2 is presented here. Unless otherwise stated, the oscillograph chart speed was 50 mm/min. Figure 8-7 shows time histories of three important independent variables following a negative "step" change in gaseous fuel rate to boiler No. 2. The change in fuel rate was about 270 lb/hr, or about 1.8% of the original value of 15,369 lb/hr. The initial rate of steam generation was 361,350 lb/hr. The boiler feed water controller was in the manual mode, as were the steam pressure and air flow rate controllers.

The pure delay time in the response of the % oxygen signal (channel 5) is approximately 1.2 minutes! This was caused largely by the time required to transport the gas sample to the analyzer. Delay time between the change in fuel flow rate and the response of superheated steam pressure was 6 seconds.

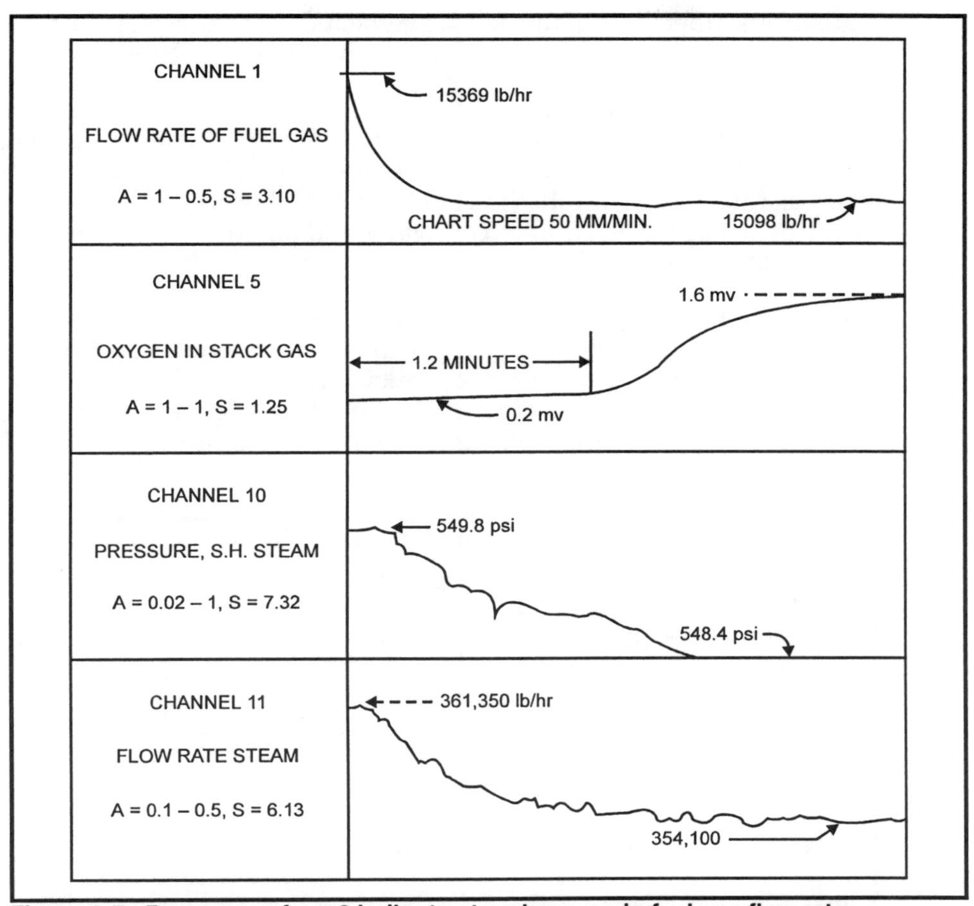

Figure 8-7. Response of no. 2 boiler to step decrease in fuel gas flow rate

Of greater significance is the measure of steam pressure (channel 10), which decreased in a roughly linear manner, with no indication of change in that trend. Therefore, the fuel rate was restored to its original value to curtail further decrease. This behavior is to be expected since, if no change in user demand occurs, the decrease in production must manifest itself as a continuous reduction in the inventory of steam in the distribution system, with consequent reduction in pressure. Systems displaying this property may be called integrating systems; they are frequently encountered among industrial processes and must be recognized when specifying controller functions or algorithms.

Outputs of the special compensating components (see Figure 8-1) showed no significant change in this or any other tests, demonstrating that these components were not functioning and, hence, did not execute the intended functions.

Numerous other tests of this kind, of which this is a sample, were executed and much additional data obtained that provides a portion of the information to be summarized later.

Figure 8-8 shows four records obtained following a positive pulse in the rate of fuel gas to boiler No. 2, with no NVR being fired. The peak value of fuel gas rate was 16,391 lb./hr, (270 lb./hr, or about 1.7%, above the initial value of 16,121 lb/hr). Boiler feed water and air supply control systems were in the manual mode.

As expected, the excess oxygen (channel 5) responded in an inverse manner, while steam production (channel 11) responded positively. The pressure of superheated steam (channel 10) rises rapidly, reaching a peak before attaining a new steady state value 1.4 psi higher than the initial value (the reason for this behavior was mentioned previously).

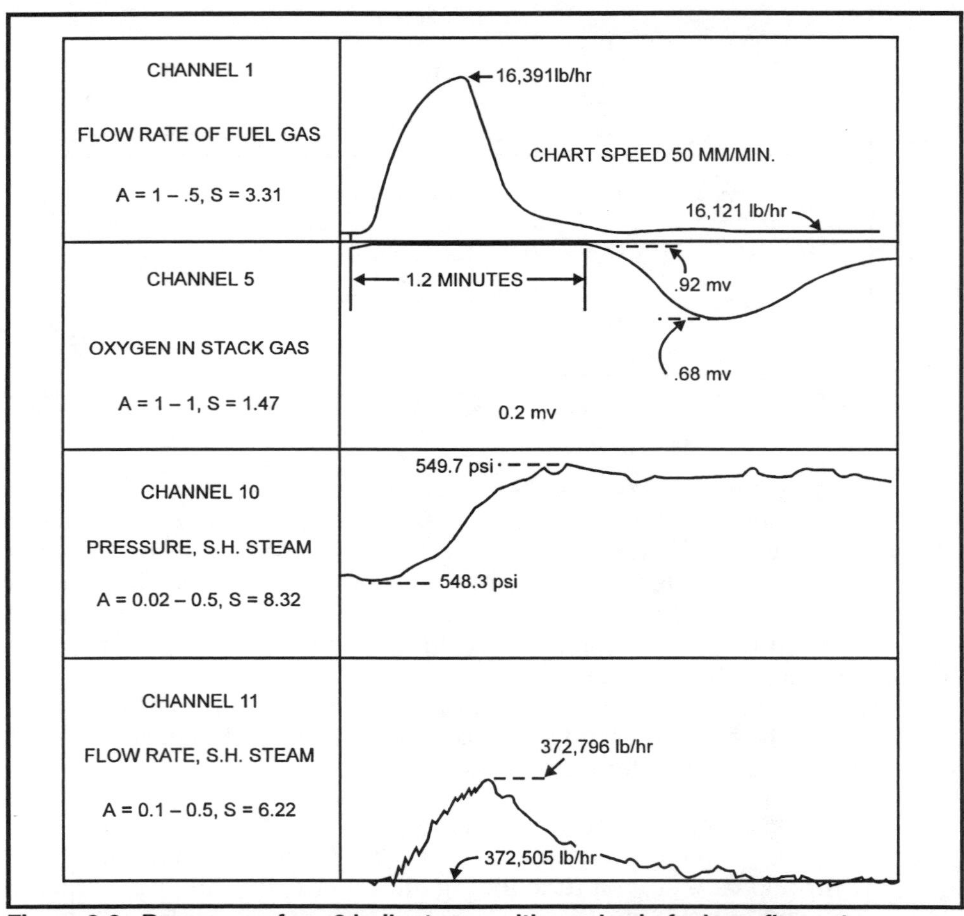

Figure 8-8. Response of no. 2 boiler to a positive pulse in fuel gas flow rate

In Figure 8-9 responses of four variables to a positive pulse change in the rate of air flow are shown. In this test no NVR was fired, and the boiler feed water and pressure control systems were in the manual mode. As will be noted, air flow rate (channel 14) follows very closely the air vane position (channel 15), the former being the primary input. The blower discharge pressure (channel 13) responds similarly.

The most noteworthy response is the excess oxygen in the flue gas (channel 5). Here, after the usual 1.2 minutes of pure delay time, the analyzer output produces a nearly symmetrical positive pulse, indicating a smooth change in O_2 from about 1.23 to 1.48%, before returning to the initial value.

Responses of four variables to a negative "step" in NVR fuel from about 12 to 9 gpm are shown in Figure 8-10. Note the characteristic negative exponential decay in steam production (channel 11), a change from 255,785 to 246,220 lb./hr. The almost linear decrease in steam pressure (channel 10) should be observed, the rate being approximately 0.42 psi/minute.

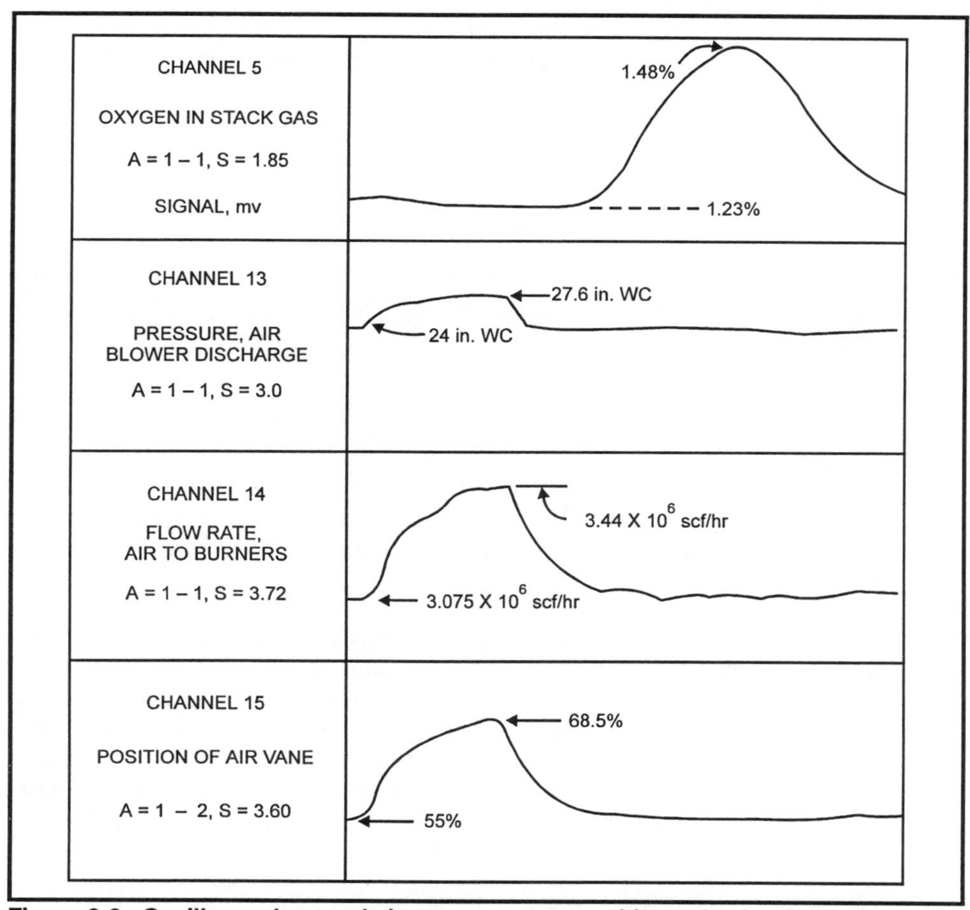

Figure 8-9. Oscillograph records in response to a positive pulse in air flow rate

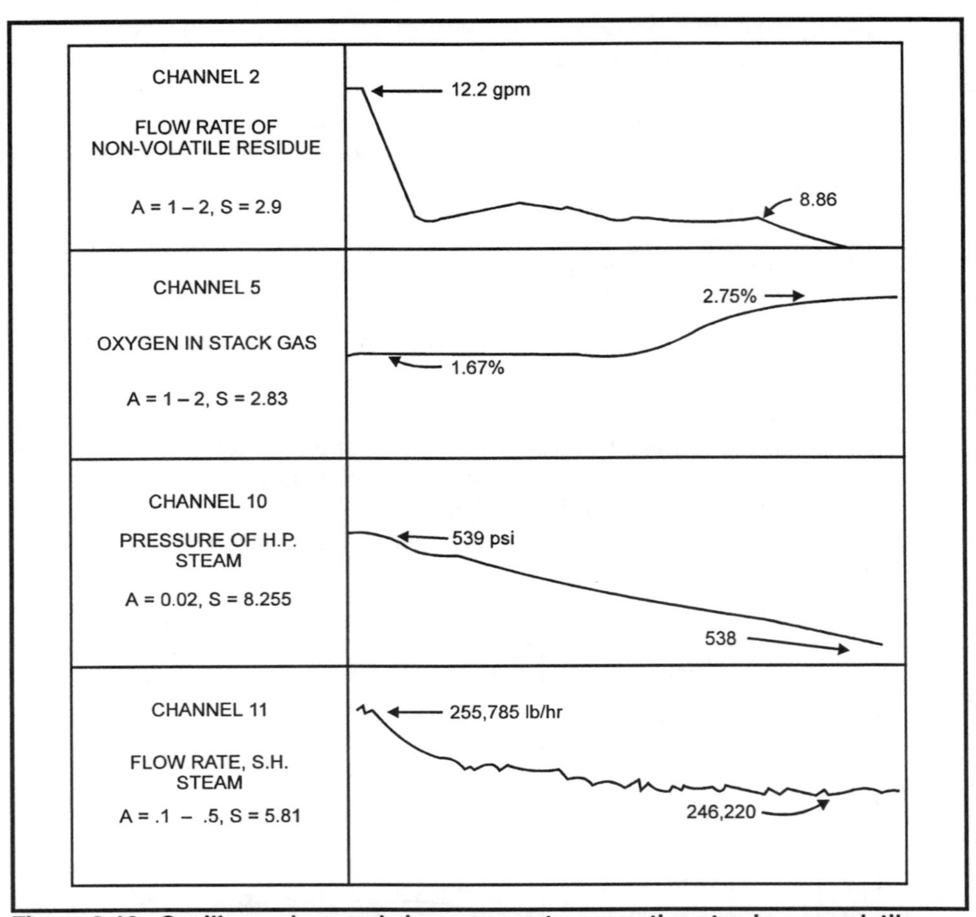

Figure 8-10. Oscillograph records in response to a negative step in non-volatile residue

Four channels of data shown in Figure 8-11 are time histories that occurred in response to a positive pulse change in NVR (channel 2), from 3500 lb./hr (7.4 gpm) to about 6500 lb./hr, and returning. All responses appeared as expected. The sensitivity of the amplifier handling the steam pressure signal was such that full scale on the chart represents a 1 psi change.

Figure 8-12 shows oscillations in the flow rate of boiler feed water (channel 9) and drum level (channel 12) during a period of steady state operation. The chart speed was 250 mm/min., 5 times the rate normally used in these tests to record dynamic response test data. The frequency of the feed water flow rate signal was roughly 7 cycles/min, that of drum level about 11 cycles/min. This performance was probably the result of using controllers imparting proportional plus integration, whereas only proportional control should be considered.

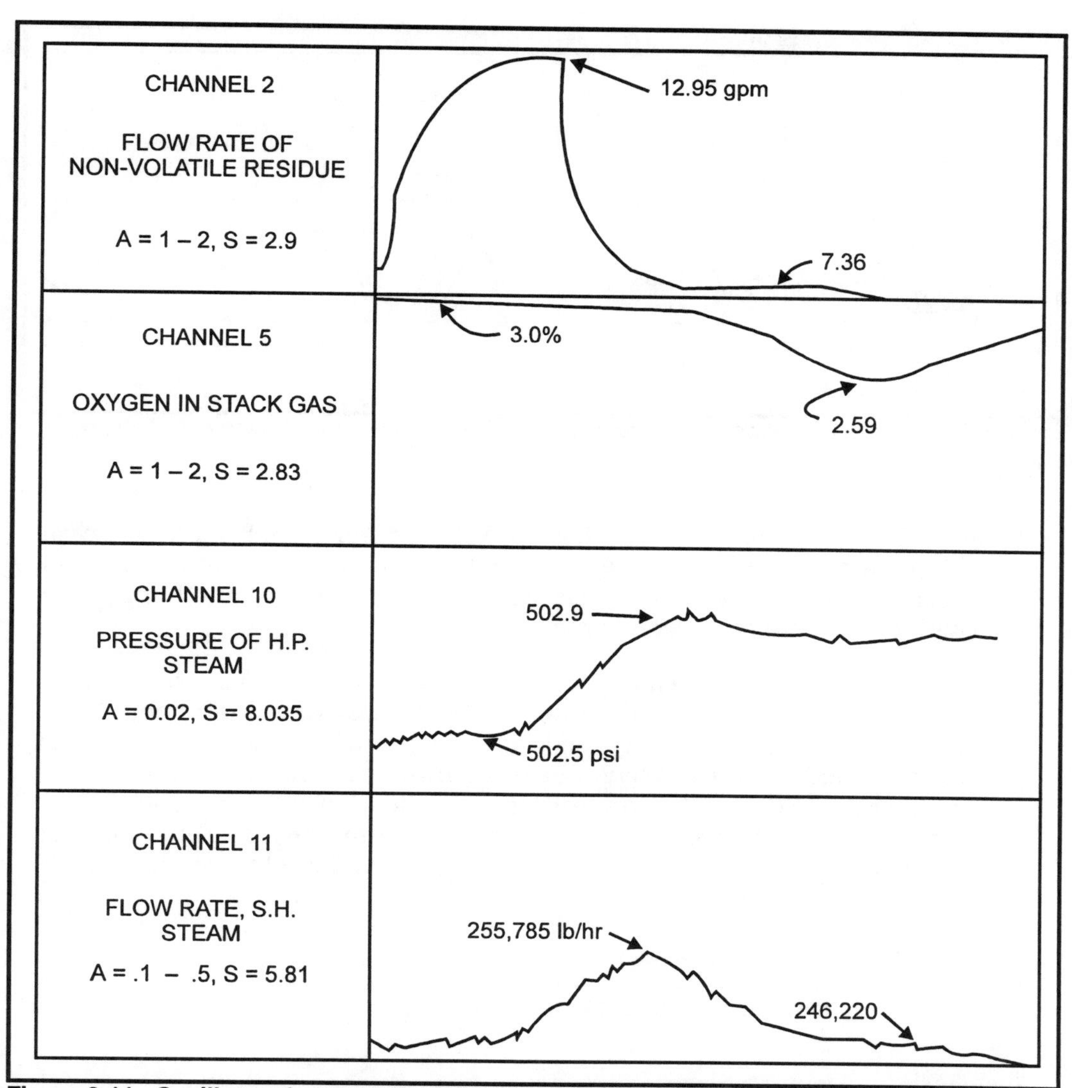

Figure 8-11. Oscillograph records in response to a positive pulse in non-volatile residue

The time histories of a positive pulse in fuel gas flow rate and the response of steam pressure from boiler No. 2 appear in Figure 8-13, along with the Bode diagram derived from those time histories. As can be seen, the pressure pulse does not close, as expected. The Bode diagram clearly shows the system can be described as a pure integrator, the amplitude being a straight line with a slope of –1, and phase angle a horizontal line at –90 degrees. Pure delay time of 6 seconds was removed from the output before processing by using the Fourier transform program.

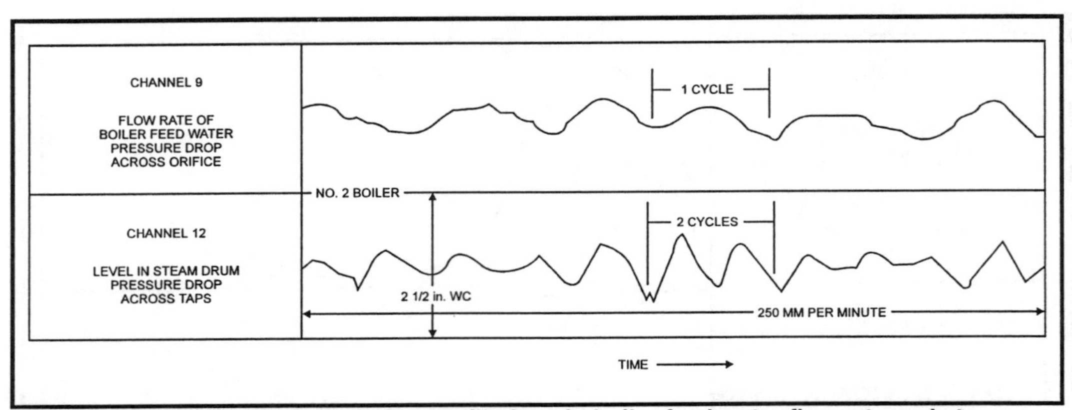

Figure 8-12. Oscillographs showing oscillations in boiler feed water flow rate and steam drum level

Results of a similar test on boiler No. 3 appear in Figure 8-14. In this case the change in fuel gas was negative, with some NVR included in the fuel. Again the frequency response shows the relation to be a pure integration (with delay time of 6 seconds removed).

The response of superheated steam flow rate to a pulse increase of fuel was also obtained, the result indicating essentially first-order behavior. Also, the response of the oxygen in the flue gas to a pulse increase in fuel was recorded. With the delay time of 1.2 minutes removed before processing the data, a "second-order" critically damped frequency response was obtained. This is characteristic of mixing and diffusion that occur in long narrow tubes and has been observed in other similar situations. This linear approximation is

$$\frac{\text{Change in } O_2}{\text{Change in Fuel Gas Rate}} = \frac{1}{\frac{s^2}{\omega_n^2} + \frac{2\zeta s}{\omega_n} + 1}$$

where $\zeta = 1$.

The delay time of 1.2 minutes (removed before data processing) was, as previously mentioned, undoubtedly due to the length of the sample line.

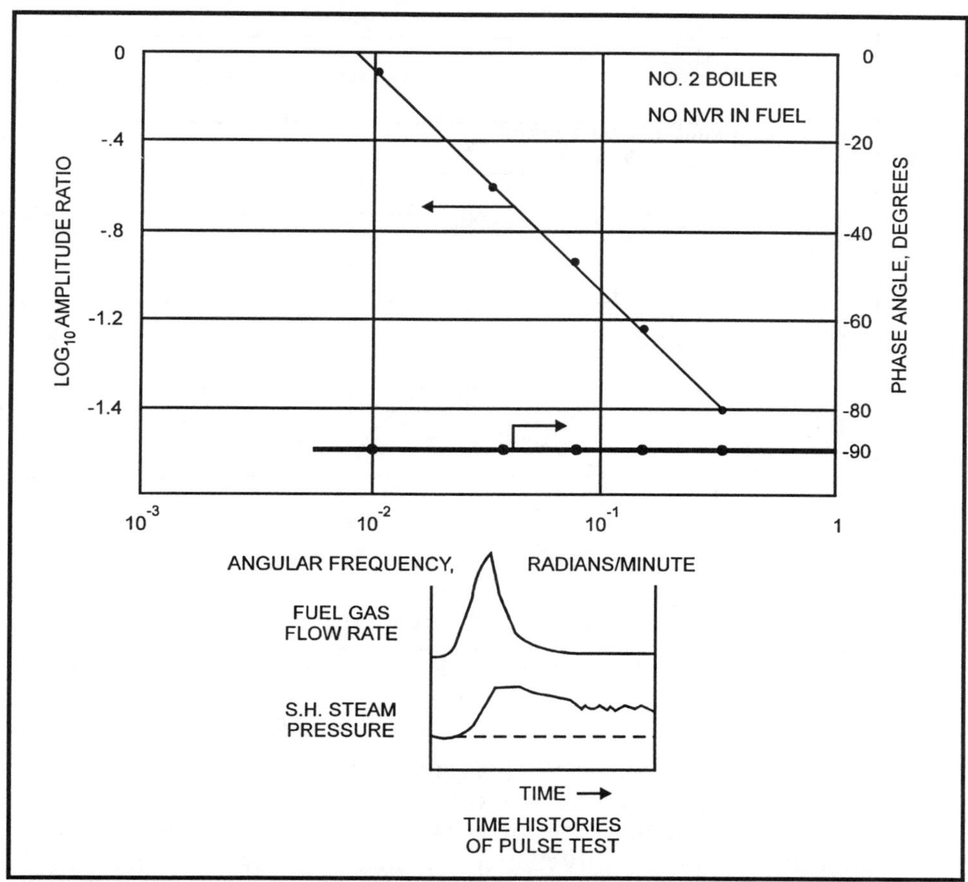

Figure 8-13. Frequency response relating pressure of superheated steam and fuel gas flow rate

Conclusions

The examples of test data and computed results presented here are sufficient to demonstrate that, with an appropriate data system, both steady-state and dynamic response data can be readily obtained from industrial processes. From such information, malfunctions of processing apparatus can be discovered, useful performance data obtained, control strategies derived, and control systems designed. In the case of the boilers considered here, the existing superheated steam pressure controller algorithm (P-I) was the major cause of poor control of this variable. Where a proportional+integral controller is used (as in this case) to control a process that imparts integration (and in addition possesses 6 seconds of delay time), two integrations appear in the denominator of the open-loop, feedforward frequency function. Thus, at zero frequency the open-loop phase angle is already –180 degrees and can become even less as frequency increases, if pure delay time is present.

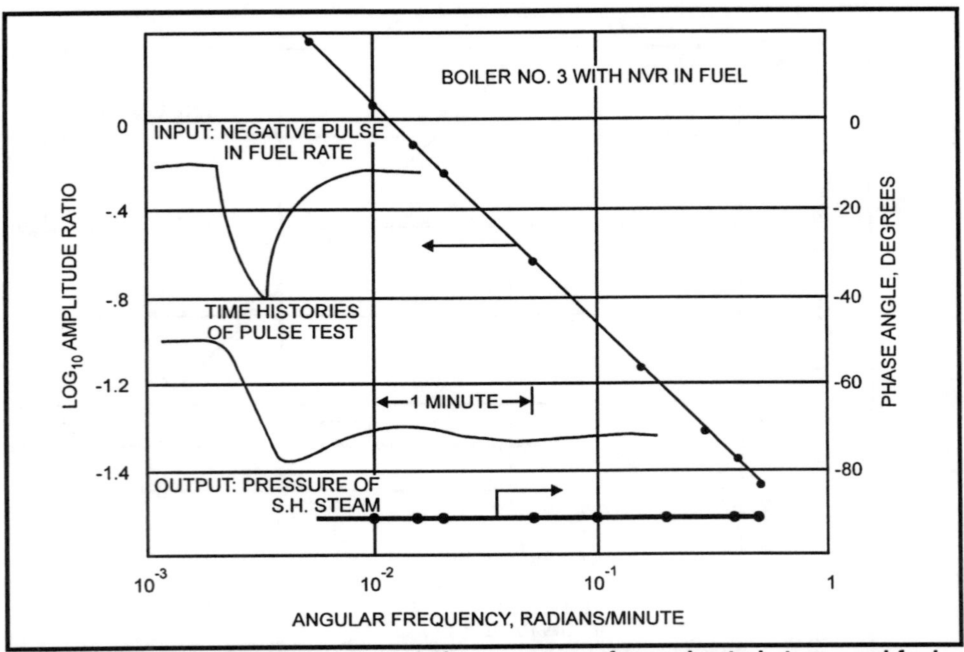

Figure 8-14. Frequency response relating pressure of superheated steam and fuel gas flow rate

Without delay time, closing the loop with a P+I controller in the feedforward path creates a second-order system, and oscillations occur if controller parameters are not carefully selected. The presence of delay increases the probability of creating an inferior system. This was responsible for the problem associated with the control of these boilers at the time of this study.

The presence of over 1 minute of pure delay time in the oxygen sampling system renders the signal from this analyzer completely useless for feedback control.

Finally, the feedback arrangement, in which the special compensating components appeared, was of no use whatsoever; these components remained essentially inert throughout this testing program.

Recommended Control Strategy and Controller Functions

A suggested control scheme is shown in Figure 8-15. The recommended algorithm for the steam pressure controller is proportional with amplification and phase lead provided by the algorithm shown below.

$$\frac{K(1 + AT_d s)}{(1 + T_d s)}$$

where A should be about 10.

This imparts a gain, which becomes 10 at high frequency, and a maximum phase lead of 55 degrees.

The purpose of this "derivative" function is to compensate for phase lag inherent in the pure delay time. Both the gain, K, and "derivative" parameter, T_d, may be found by trial and error. However, if the feedforward gain of all components, including the process, and the pure delay time are known, optimum parameters may be found from the software, PCPARSEL[1]. Table 8-3 presents these control parameters and resulting crossover frequencies derived from PCPARSEL, for both purely proportional and proportional with "derivative" controllers. Note that process gain has been taken as unity and the process integration parameter as 100. Various values of delay time have been assumed in order to show the deleterious effect this has on the closed-loop performance.

The boiler water level controller should provide proportional action only since that process is also a pure integrator.

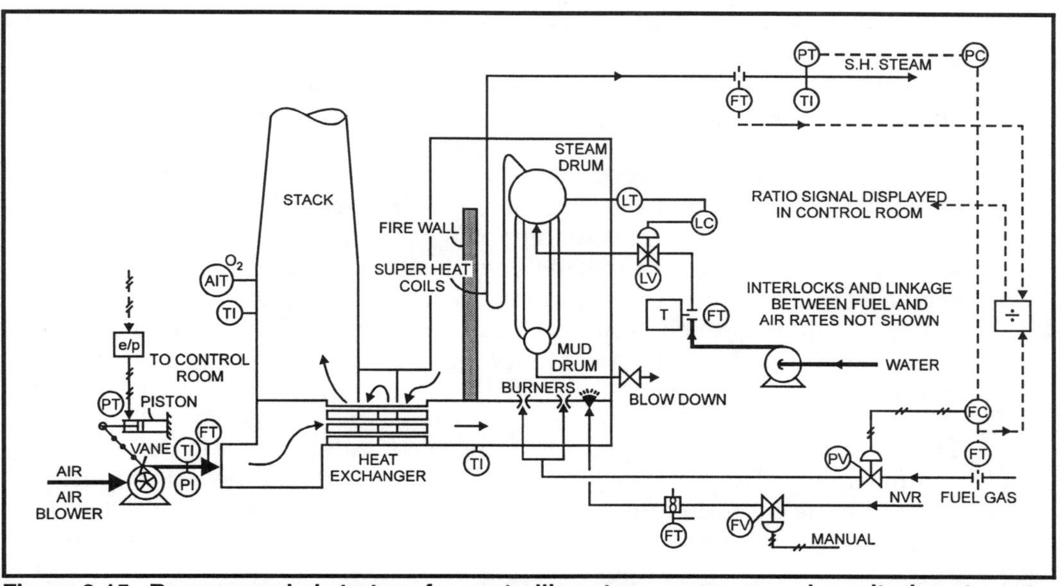

Figure 8-15. Recommended strategy for controlling steam pressure and monitoring steam to fuel gas ratio

1. PCPARSEL is a software program developed by the author but no longer available from ISA. See Appendix A.

Table 8-3. Optimum Controller Parameters for an Integrating Process with Two Controller Algorithms

Process gain	Process integration	Delay time, sec. DT	Total gain (closed loop)	Controller "derivative" parameter	Crossover frequency, radians/min
PROPORTIONAL					
1	100	1	67.6	0	1.35
1	100	2	33.8	0	0.68
1	100	3	22.5	0	0.45
1	100	5	13.5	0	0.27
1	100	6	11.3	0	0.23
1	100	10	6.8	0	0.14
PROPORTIONAL PLUS "DERIVATIVE"					
1	100	1	87	0.045	2.46
1	100	2	43.5	0.090	1.23
1	100	3	29.0	0.135	0.82
1	100	5	17.0	0.225	0.49
1	100	6	14.5	0.270	0.41
1	100	10	8.7	0.45	0.25

Note the improvement in crossover frequency obtained by introducing the "derivative" component in the controller function or algorithm.

The delay time in the oxygen sampling system must be reduced by decreasing the length of the sample line or by increasing sampling velocity, or both, if the analyzer signal is used in a closed-loop control scheme. The response of the analyzer itself may probably never be satisfactory for on-line control in this instance. In any event, a display of the record of oxygen content in the stack gas should be available in the control room, and efforts made to maintain excess oxygen at a reasonable minimum.

The control system for regulating boiler feed water, utilizing the two compensating components shown in Figure 8-1, might have provided some benefit, provided these components were operable. The strategy has been applied, with gratifying results, on numerous boiler installations.

Presentation of accurate measures of the flow rates of steam and fuel, especially fuel gas, would encourage economy. If a sensitive display of the ratio of steam production to fuel consumed were conveniently available in the control room, plant operators would be motivated to maintain this ratio at its maximum, consistent with acceptable minimum excess air.

9

Production of Aluminum Fluoride

In this chapter we describe a study of a unique process, that by which aluminum fluoride is prepared, a necessary component in the production of aluminum. Considerable engineering judgment and ingenuity was exercised in the construction of the facilities and in subsequent operation of the process. The pneumatic control systems that existed most certainly had been selected and installed with very little knowledge of process behavior and without any dynamic information whatsoever. One cannot fail to be favorably impressed by the originality and skill required to conceive, construct, and operate a process such as this.

The reader should note that even processes that appear large, unwieldy, and ponderous are indeed quite sensitive and reproducible in their behavior, but that data systems of considerable capability are needed to obtain satisfactory information about either steady-state or dynamic behavior.

In describing this plant study, the major task that confronted the author was abstracting the essence of the work from more than 100 pages of text, diagrams, oscillograph records, and extensive tabulations of data. Presenting the results in easily digestible form agreeable to an audience having little interest in the particular process, but who may be seeking assistance in solving problems associated with processes of an entirely different nature, was another formidable and additional challenge. We have tried to keep the reader in mind in preparing this resumé.

The Process

The process to be considered here, and shown in Figure 9-1, was carried out in two parts: first the production of hydrogen fluoride (HF), then the reaction of this gas with aluminum hydrate ($Al_2O_3 \cdot 3H_2O$) to produce the desired product, aluminum fluoride (AlF_3). The HF gas was produced by the reaction of oleum with feldspar in a rotating kiln about 40 feet long and 8 feet in diameter, heated externally. A belt conveyor supplied the crushed feldspar to the screw feeder, or premixer, which delivered it to the kiln in which the following reaction occurred:

$$CaF_2 + H_2SO_4 \longrightarrow 2\,HF(g) + CaSO_4$$

The theoretical yield is 0.512 lb. of HF per pound of pure CaF_2, or 0.496 lb. of HF per pound of spar (97% CaF_2). The calcium sulfate, continuously discharged from the opposite end of the kiln, was slurried with water and directed to disposal.

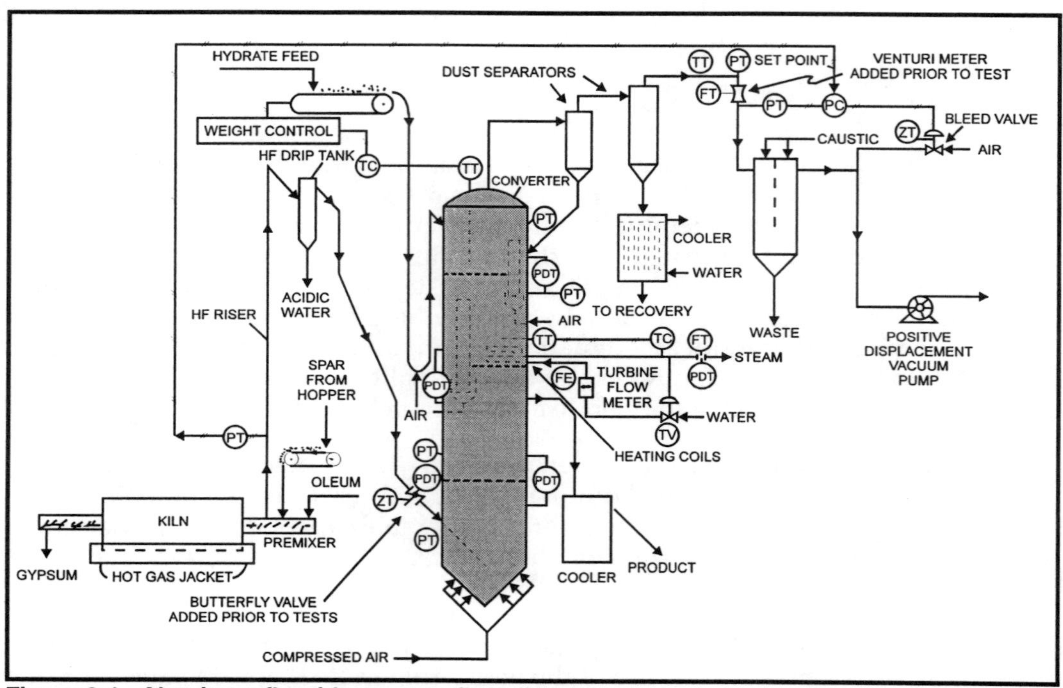

Figure 9-1. Aluminum fluoride process flow diagram

The HF gas, after passing through a "drip tank" and butterfly valve (installed prior to the testing program) entered the bottom of the reactor. The reactor, or converter, a cylindrical vessel about 6 feet in diameter and 40 feet high, was supplied at the top with finely divided $Al_2O_3 \cdot 3H_2O$ by means of a belt conveyor. Air admitted at various places, along with the HF vapor, served to keep the solids within the converter in a state of fluidization.

The converter was divided into three sections, the upper two contained down-comers from each perforated plate. The lower plate contained only perforations through which the rising gas flowed, reacting with the descending finely divided solids, which entered above the top-most plate.

Product was withdrawn at a point somewhat above the bottom distribution plate. Of considerable importance were the dust separators that removed finely divided solid material from the reactor overhead. Solids collected in the first separator returned to the top of the reactor, that from the second were cooled and recovered. The effluent gas, after passing through a Venturi meter and caustic scrubber, was withdrawn by a Nash vacuum pump. In order to avoid discharge of HF gas into the atmosphere, the kiln was operated at a slight vacuum.

Within the converter the following reactions occur:

- Dehydration of $Al_2O_3 \cdot 3\ H_2O$ (endothermic), and
- Reaction with HF in the lower part (exothermic).

Per 100 lbs. of hydrate ($Al_2O_3 \cdot 3H_2O$), the following are involved for complete reaction:

HF required for complete reaction —— 77 lb.

Water formed —————————— 69 lb.

AlF_3 formed —————————— 108 lb.

The gaseous HF from the premixer was conducted upward through a water-cooled jacketed steel pipe and a liquid separator; HF then flowed downward to enter the cone-shaped bottom section of the converter. The vapor at this point made contact with finely divided solid material, probably largely AlF_3. The vapor, distributed beneath the lower perforated plate, was assumed to produce a uniformly fluidized mixture above the bottom plate (bed 1).

In bed 1 further preheating, and possibly some reaction, takes place. From above this bed solid product was removed through a conduit leading to the cooler. This cooler was an ingenious device through which the solid product flowed in a fluidized manner, contacting water-cooled surfaces. Gases from bed 1 passed through bed 2 in which the majority of reaction probably occurred.

Fresh hydrate ($Al_2O_3 \cdot 3H_2O$), supplied by a belt conveyor, entered through a fluidized standpipe into bed 3 (top), where heating and dehydration occurred. Dehydration is necessary because reaction with HF does not occur at an acceptable rate with hydrated alumina. Since the thermal energy required for dehydration is less than the heat liberated in the reaction of HF with Al_2O_3, the removal of heat is necessary through the cooling coils installed at this point.

The major down-flow of solids in the converter was through the vertical downcomers from the two upper plates. To ensure fluidization in these downcomers, air was admitted into the apex of each cone into which the lower end of each downcomer terminated. (This is shown in Figure 9-1.)

In an ideal state of fluidization each bed is visualized as composed of a freely-flowing, well-agitated, uniform mixture of gas and finely divided solids, acting almost like a liquid. Circulation of solids, with both upward and downward movement, apparently occurs. When operating properly the inventory of solids in each section remains fairly constant; and carry-over into the overhead cyclones is not sufficient to create congestion. The terminal vacuum pump draws gases and vapor from the converter through the cyclone separators and caustic soda scrubbers.

Evolution of HF gas into the atmosphere was prevented by operating the converter at a pressure less than atmospheric. Since it was impossible to maintain a vapor-tight system, encroachment of air at several points could not be prevented. As mentioned previously, air was deliberately introduced to assist in creating fluidization. For these reasons air comprised a significant fraction of the gas and vapor leaving the converter.

The presence of air decreases both the rates of reaction in the converter and condensation within the scrubber and adds to the load on the terminal vacuum pump. Hence, it was desired that the admission of air be kept to a minimum consistent with acceptable fluidization. A minimum of carryover of solids to the cyclone separators was also desirable.

Existing Control System

The original control system consisted of two temperature control loops: one to regulate the temperature above the top plate of the converter by varying the flow rate of hydrate feed, the other for control of the temperature above the central plate in the converter by varying the flow rate of water to cooling coils located in the converter. In addition, a control system existed for regulation of pressure near the effluent of the kiln associated with the system used to control the pressure just prior to the effluent gas scrubber. The latter was composed of a cascaded control arrangement with the output of the kiln pressure controller serving as the setpoint of the separator effluent (or scrubber influent) pressure controller. (These arrangements are shown in Figure 9-1.) Regulation of the flow rates of both hydrate and spar depended upon the control systems associated with the belt conveyers. These systems, as well as the method for adding oleum, left much to be desired. Considering the state of the art at the time of this study, the components for all pneumatic control systems had probably been selected in a rather arbitrary manner and installed without the knowledge or information needed to produce designed systems.

The Data System

The data system used for conducting the test program consisted of 12 channels of Brush recording oscillographs provided with amplifiers having a maximum gain of 1 microvolt per division of pen deflection and with the ability to follow signals faithfully up to 125 Hz. Strain-gage-type transducers were used to measure pressures and pressure gradients. The lowest pressures were measured with Statham unbonded gages with a range of ±1 psi and a frequency response flat to 100 Hz. Calibrated suppression circuitry enabled shifting the zero of each channel to an arbitrary, but known, position so that gain sufficient to observe very small changes in the measured variable could be employed. This feature is essential in order to record small deviations, especially when performing dynamic tests. Higher pressures and pressure differentials were measured with Viatran® strain gage transducers with appropriate ranges. The bridges of all pressure transducers were excited with 10 volts dc giving a full-scale output of about 5 mv per volt of excitation For temperature measurements high-gain dc amplifiers were required, along with auxiliary circuits associated with the resistance elements. Several industrial thermocouples were in service, but, since these were not calibrated, the information obtained was suspect. Positions of valve stems were measured with linear variable differential transformers (made by Schaevitz) excited from a 2400 Hz voltage source. Demodulation was carried out in the same component. The output signals were large enough to enter directly into the recorder amplifiers. Displacements as small as 1/100 in. WC could be detected. Pressure and differential pressure transducers were located in a protected area and connected to the process through 1/4-inch PVC tubing. To avoid damage to the transducers, the leads connecting transducer to process were purged with nitrogen at a very low pressure. During periods of testing the purges were not used.

Pre-test Alterations and Calibrations

As mentioned previously, a butterfly valve was installed in the HF influent line ahead of the converter. In addition, a Venturi tube was inserted between the converter overhead dust separators and scrubbers. These two items were calibrated using a standard orifice properly located in a testing facility. A Nash vacuum pump was used to draw air through the calibration assembly, shown in Figure 9-2. The dimensions of the flow meters are given in Table 9-1, and the results of the calibrations follow.

Table 9-1. Dimensions of Meters

Meter	Throat Diameter, inches	Pipe I.D., inches	d/D
Orifice*	4.048	10.02	0.404
Venturi	5.000	10.02	—

* Flange taps

For the Venturi flow meter:

$$W_{vm} = 9700\sqrt{\Delta P_{vm}\rho_{vm}}$$

where

W_{vm} = Flow rate through Venturi, lb./hr
ΔP_{vm} = Pressure drop across Venturi, inches of water
ρ_{vm} = Density of gas measured at upstream conditions, lb./ft

For the butterfly valve:

$$W_{bv} = Ke^{kz}\sqrt{\Delta P_{bv}\rho_{bv}}$$

where

W_{bv} = mass rate of flow, lb./hr
ΔP_{bv} = pressure drop across butterfly valve
ρ_{bv} = density of fluid at upstream conditions
z = position or displacement of valve vane, % open
e = base of natural logarithms
K = constant depending upon size of valve
k = constant depending upon geometry of valve vane and throat

The above form has been found to describe the performance of a large number of valves of different sizes, types, and configurations. To calibrate a valve it is necessary to measure the stem position, z, the pressure gradient, ΔP, and to have an independent measurement of the mass rate of flow, W. Well-known graphical procedures may be used to find the constants K and k. For the valve in service in this study, the relation was

$$W = 1200e^{0.0263z}\sqrt{\Delta P_{bv}\rho_{bv}}$$

The range in flow rates for this valve is given as

10% open—300 cu. ft./min.

70% open—1500 cu. ft./min of air at Standard Conditions (14.7 psia, 32°F)

A calibrated turbine flow meter was used to measure the flow rate of water to the cooling coils installed above bed 3, and an orifice plate to indicate the flow rate of steam from these coils.

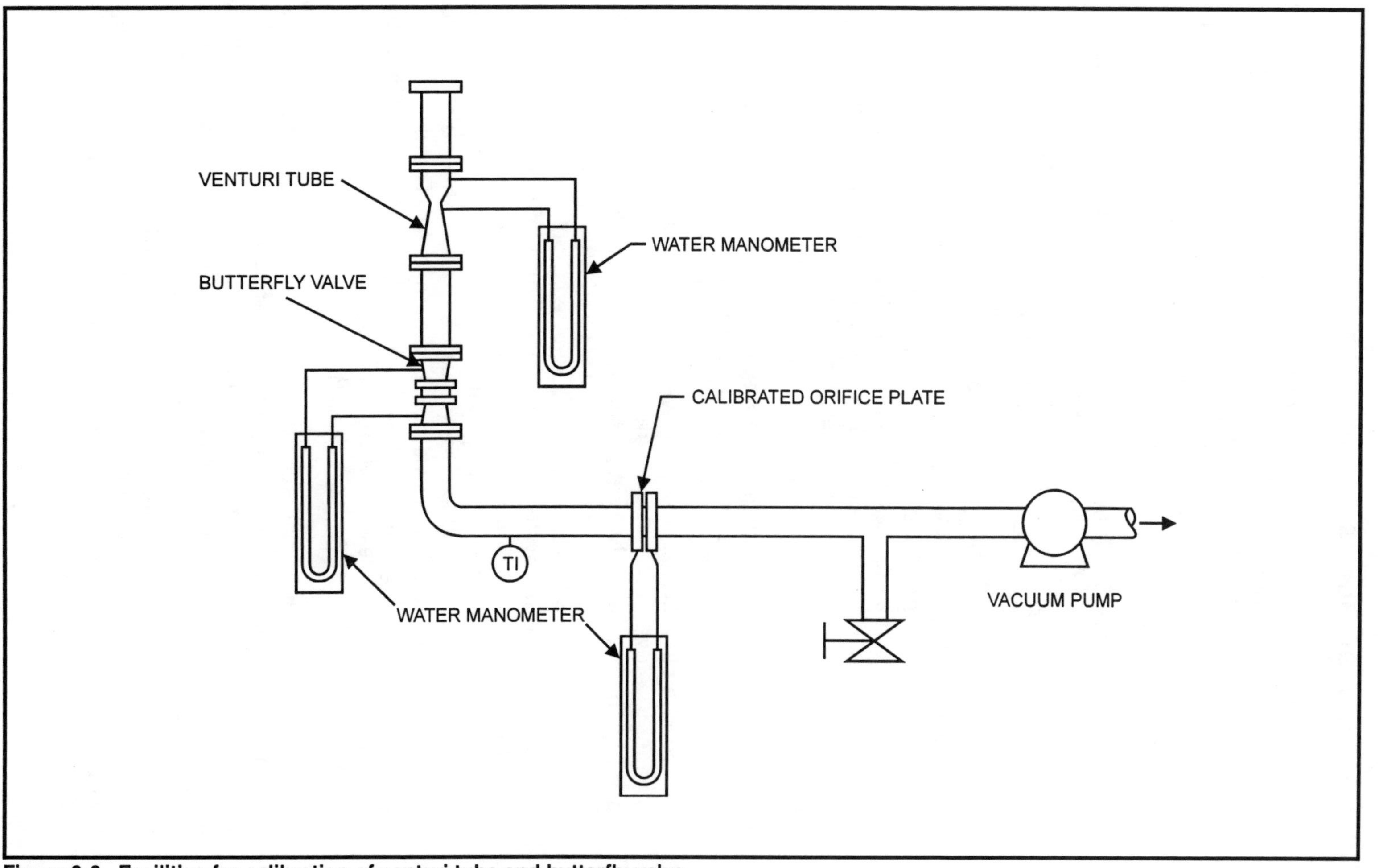

Figure 9-2. Facilities for calibration of venturi tube and butterfly valve

Preliminary Observations

A persistent processing difficulty plaguing the operation of this process was the apparent periodic overloading of the space above bed 3 (top) with solids. These recurring episodes were referred to as "dumping." Oscillograph records, obtained just prior to and during a typical "dump," are shown in Figure 9-3. The sequence of events and a possible explanation of causes follows.

Referring to Figure 9-3, the first, and possibly insignificant, discernible changes are a very slight increase in pressure at the Venturi, a similar diminution in the pressure drop across bed 2, and possibly a minute increase in the pressure drop across bed 3. Then there is a sharp decrease in pressure at the Venturi, followed by an abrupt rise in pressure at the premixer. Apparently, to restore the pressure at the kiln and premixer, the control system closes the inlet butterfly valve, thereby diminishing the flow rate at that point. Meanwhile, the pressure drop across bed 2 slowly increases for about 2 minutes and finally is restored to its original value after 20 minutes. During this period the pressure gradient across bed 3 (top) slowly diminishes for 5 minutes, then slowly recovers in about 15 minutes.

This set of records dramatically demonstrated that the existing control systems were not very effective in restoring pressures in the converter. But more importantly a likely source of the "dumping" action was indicated, *i.e.*, a recycling of material carried upward, accumulating, and periodically returning to a lower level. A very likely source of this behavior could indeed be the overhead dust separator. Material too light to remain in the converter but too heavy to leave the first separator could accumulate. Possibly the first separator became so congested that flow through the overhead gas-handling system was restricted. This would explain why the sharp diminution in pressure at the Venturi was the first major event to be observed in this sequence. From the time histories of pressure drops across beds 2 and 3, it appears that excess solids descended rather quickly to bed 2, causing the pressure drop across this bed to increase. The slowly diminishing pressure gradient across the upper bed (bed 3) may indicate that solids continue to descend therefrom, further aggravating the condition on bed 2.

The temperature of material on bed 3 showed very little change, however, the temperature in bed 2 slowly diminished before returning to its initial value. The latter observation may provide additional evidence that solids at a lower temperature have entered the converter, and that this material could have originated from the dust separator. (It should be mentioned that this temperature measurement was obtained with an industrial thermocouple (TC), and could be considerably in error owing to sensor location, the mass of the industrial type sensors with their sheaths, radiation and conduction errors, etc.).

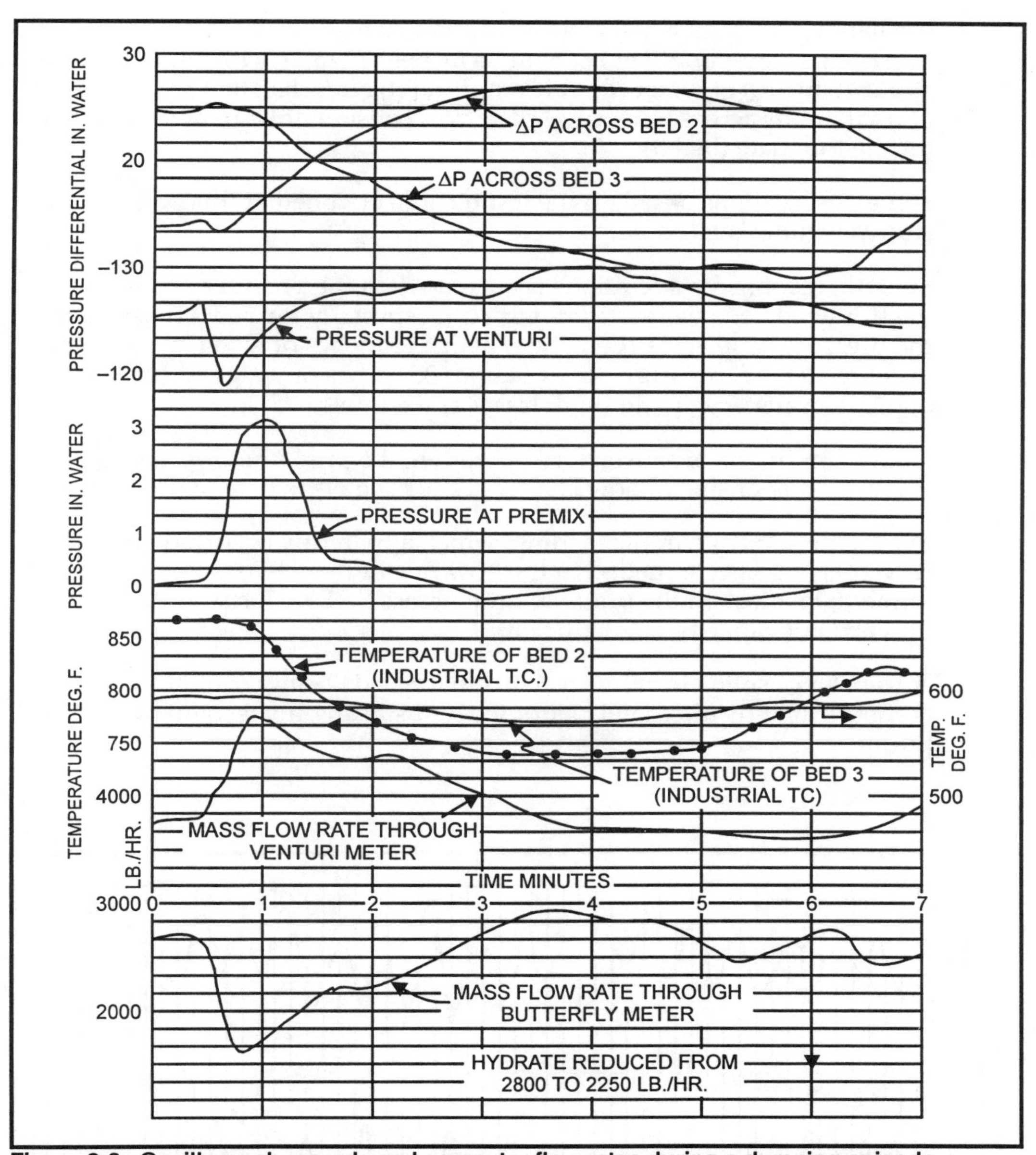

Figure 9-3. Oscillograph records and computer flow rates during a dumping episode

It is interesting to note that during this transient period, the increase in total mass flow rate through the Venturi appeared to approximate the diminution of mass flow rate through the butterfly valve.

Conditions within the reactor were probably never at a steady state. This was shown by the records obtained during a period of operations considered "normal" (presented in Figure 9-4). The pressure drop across bed 3 (top) averages about 12 inches of water with double amplitude oscillations of about 6 to 8 inches of water. Across bed 2 (center) the average pressure gradient is about 35 inches of water with oscillations of about 15 inches of water (double amplitude). Note also that the frequency of oscillations across the central plate was considerably greater than that for the top plate (bed 3).

Some observations based on the test records obtained at this point in the study are summarized here.

Smooth, responsive operations occurred when the pressure drops across the fluidized beds were moderately low, especially for bed 3 (top). When the pressure drop across bed 3 was about 12 inches of water, performance was excellent. A pressure drop of about 35 inches of water across bed 2 appeared satisfactory, although less was desirable.

Under conditions of satisfactory operation, the pressure drop across bed 3 was very responsive to flow changes occurring elsewhere in the system.

During periods of chronic loading, which apparently occurred because of recycling solids, a reduction in the flow rate of fresh hydrate produced no immediate effect on the pressure drop across Bed 3. This suggests that downcomer capacity was sufficient.

A transfer of solids from the top bed to the beds below could be accomplished by alternately stopping and starting the terminal vacuum pump.

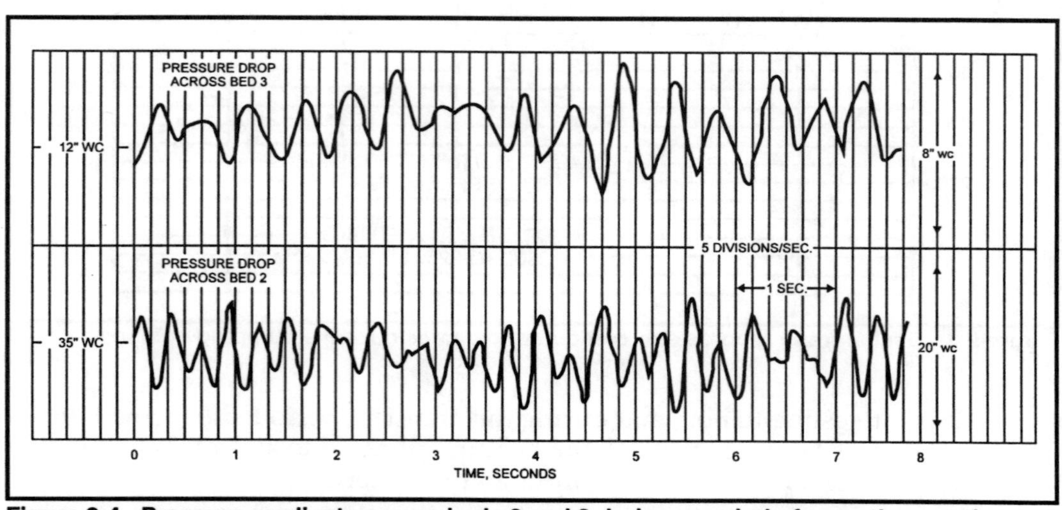

Figure 9-4. Pressure gradients across beds 2 and 3 during a period of smooth operation

In the course of this study the presence of a high-frequency pressure oscillation within the converter was observed. To investigate this further, oscillograph records were obtained using a paper speed of 125 mm/sec (a record of about 2 seconds is shown in Figure 9-5). These have been arranged to suggest that this observed disturbance originated from below bed 1 and progressed forward to the top of the reactor and upstream back to the premixer. This speculation is based on the irregularity of the records encircled in Figure 9-5. Here the pressure changes at the butterfly valve appear to occur after those below bed 1. Careful consideration of these records led to the conclusion that the phenomenon was probably the result of unsteady flow rate of gas through the lower plate.

The principal purpose of presenting these records is to emphasize that very sensitive, and very high speed, measurements may be required when investigating even seemingly ponderous and massive processing systems. Otherwise, information, if useful only for diagnostic purposes, may not be recovered.

Control System Studies

The control system for regulation of the premix pressure was suspected as being inadequate. The reason for this conclusion was the relatively large volume between the point at which corrective action takes place (bleed valve) and the point of pressure sensing (in the riser from the kiln). The presence of a "snubber" in the line, between the pressure-sensing point and the controller in the control room, was also suspected as contributing to poor control. In addition, following almost any disturbance, a period of slowly attenuating oscillations of premix pressure was observed on numerous occasions. (Figure 9-6 shows a typical set of such records.) Only three measured signals are shown in the figure:

1. Displacement of the bleed valve stem, shown as % open (lower record).
2. Pressure at inlet of Venturi meter (next to lower record).
3. Premix pressure, in. water vacuum (top record).

The three remaining records have been computed as:

a. Flow rate through the butterfly valve

$$W = 1200e^{0.0263z_{bv}}\sqrt{\Delta P_{bv}\rho_{bv}}, \text{ lb/hr}$$

where z_{bv}, the position of the butterfly valve, was set at 57.2% open, and the density of HF gas was computed at an assumed temperature of 200°F. Hence,

$$W = 5360\sqrt{\Delta P_{bv}\rho_{bv}}$$

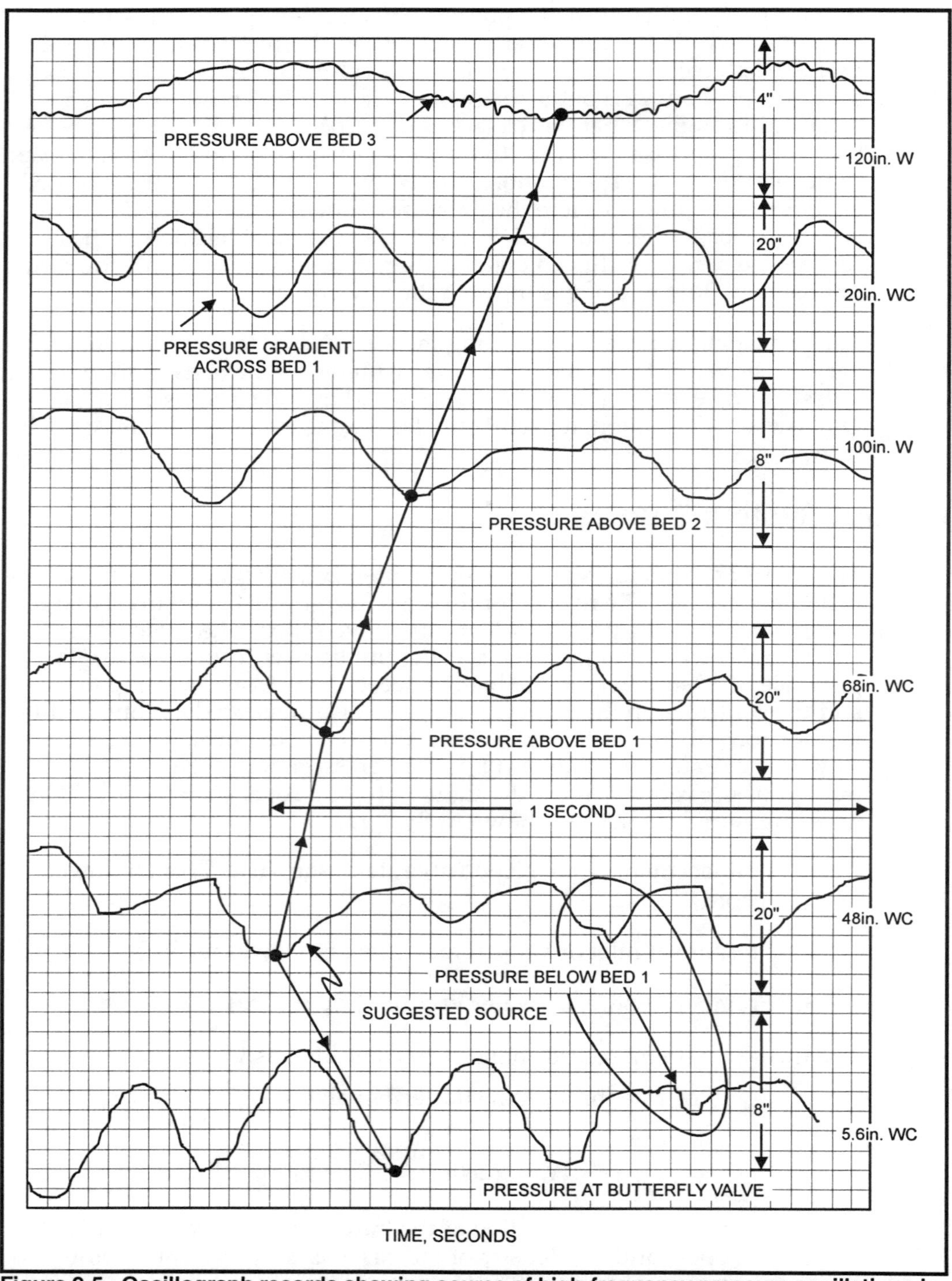

Figure 9-5. Oscillograph records showing source of high-frequency pressure oscillations in reactor

b. Flow rate through the Venturi meter (third from top), computed from

$$W = 9700\sqrt{\Delta P_v \rho_v}, \text{ lb/hr}$$

c. Theoretical flow rate through the Venturi meter (fourth from top). This was calculated assuming that only HF passed through the butterfly valve and reacted completely with $Al_2O_3 \cdot 3H_2O$ and that no air enters the system at any point.

Several interesting observations can be made from a study of these records and computations, but these may have only limited significance to the casual reader. Suffice only to observe the poor performance of the existing pressure control system.

Dynamic Tests of Control Components

During plant shutdown the existing control system was tested. An abrupt pulse input in pressure was introduced to the pressure sensor located near the kiln, and the outputs from each component in the control system measured. (The records of a typical pulse test are shown in Figure 9-7.) The progressive magnitude of pure delay time along the open-loop path is obvious, increasing from about 0.2 seconds at the kiln effluent pressure transmitter to 2.3 seconds at the terminal valve. These data show that the two components responsible for the major part of the pure delay time are the two pneumatic controllers, the first contributing 0.2 and the second, including the transmission line, 2 seconds. Even the lesser value is sufficient to contribute to the poor performance of any negative feedback, pressure control system used between the premix pressure and the bleed valve of this process.

Frequency responses were derived from the time history data of Figure 9-7, using the computer routine yielding the trapezoidal approximation of the Fourier transforms of the individual records and the ratio of the Fourier transforms for the signal pairs of interest. Figure 9-8 shows the derived frequency response of the kiln pressure sensor and transmission line, including the snubber, up to controller No. 1, located in the control room. Figure 9-9 gives the results for the entire system, from kiln pressure sensor input to bleed-valve stem position. The performance functions for these systems are shown below in Table 9-2.

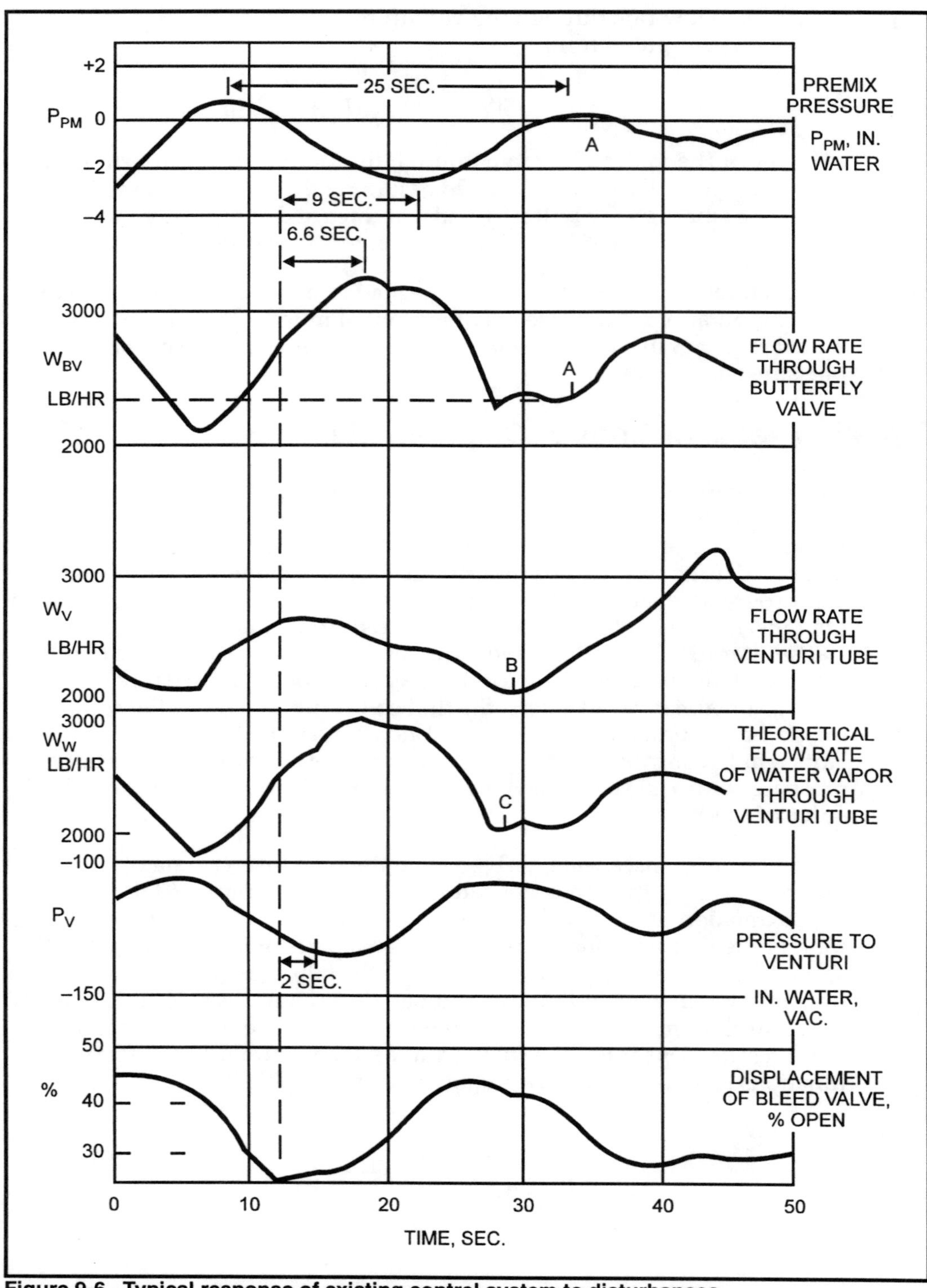

Figure 9-6. Typical response of existing control system to disturbances

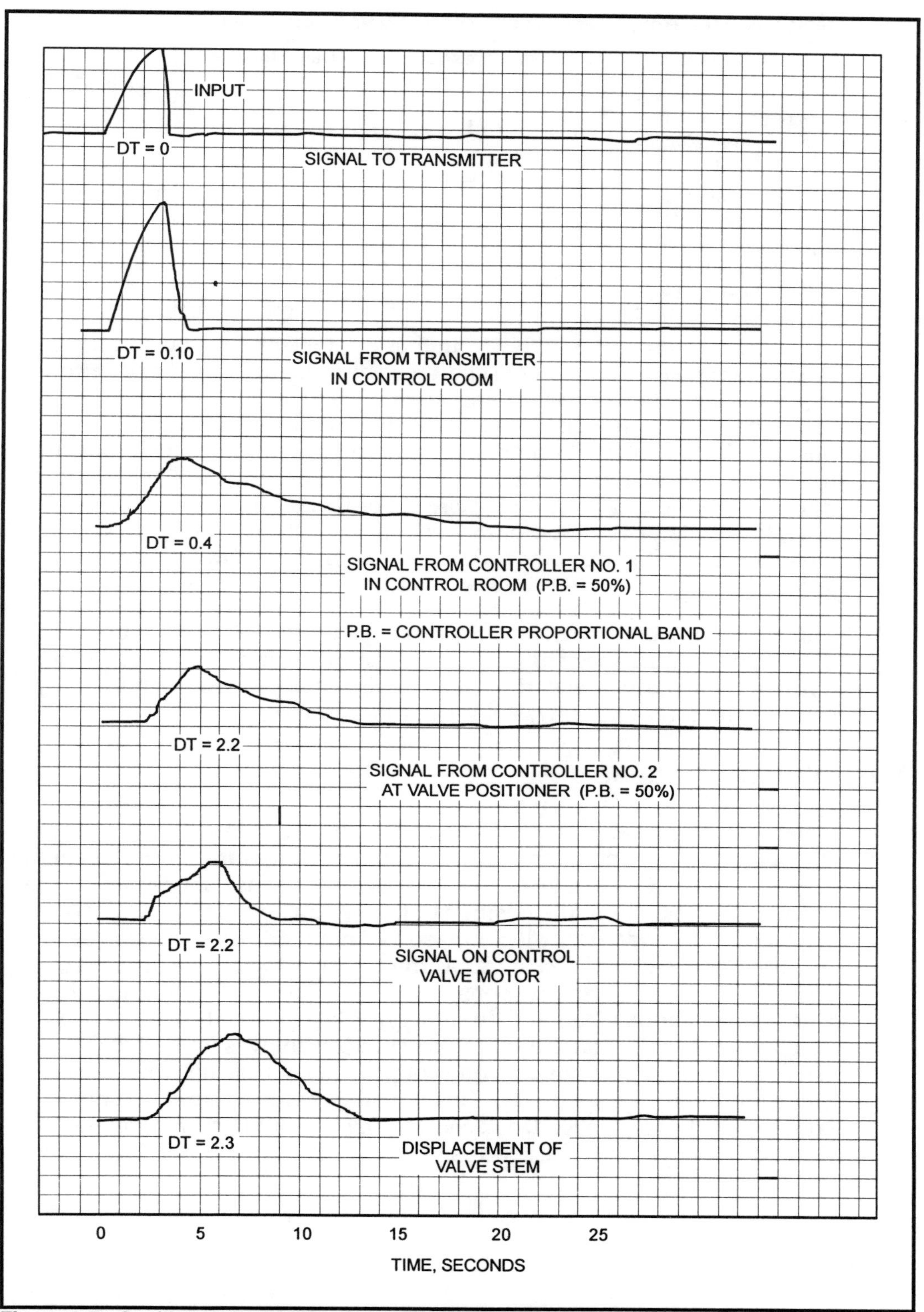

Figure 9-7. Oscillograph records of dynamic test of existing control components

Table 9-2. Frequency Response of Existing Control Components

Component	Frequency response	Delay time, sec.	
Pressure sensor and transmission line to controller No. 1	$\dfrac{0.645}{(1+0.33s)\,(1+0.22s)}$	0.1	(See Figure 9-8)
Pressure sensor, controller No. 1, controller No. 2 to bleed-valve displacement	$\dfrac{-8.05}{(1+2.5s)}$	2.3	(See Figure 9-9)

One may be tempted to derive from these test records the performance functions for each component separately, by using the input and output signals associated with each component. Except for the first two components, the pressure sensor-transmitter and controller No. 1, this is possible. But beyond that point the input to a given component (output from the preceding component) may not be sufficiently rapid to excite the dynamics of the component of interest. For this reason, except for delay time, the dynamic response of the remaining parts of the control system were not computed from these data. However, the overall dynamics shown in Figure 9-9 are considered valid.

This rather detailed analysis of the existing control system was made to document performance and to locate components having especially poor dynamics, including delay time. The significant delay time exhibited by some of the components was of greater importance.

Dynamics of Process

Several tests were also made to obtain dynamic response data on parts of the plant while in operation, but with converter pressure controllers in the manual mode. Of particular interest was the response of pressure at the premixer to changes in the position of the bleed valve. To execute this test, the bleed valve was stroked in a pulse-like manner, and the response recorded of pressures, back as far as the pressure tap on the premixer. Figure 9-10 shows typical responses when the bleed valve, initially closed, was opened about 40% and then rather quickly returned to the closed position. The major cause of poor pressure control was immediately obvious. There was about 4 seconds of pure delay time between the bleed valve and premix pressure tap. This time delay was so large that no conventional negative feedback control system could accomplish the desired objective with this process.

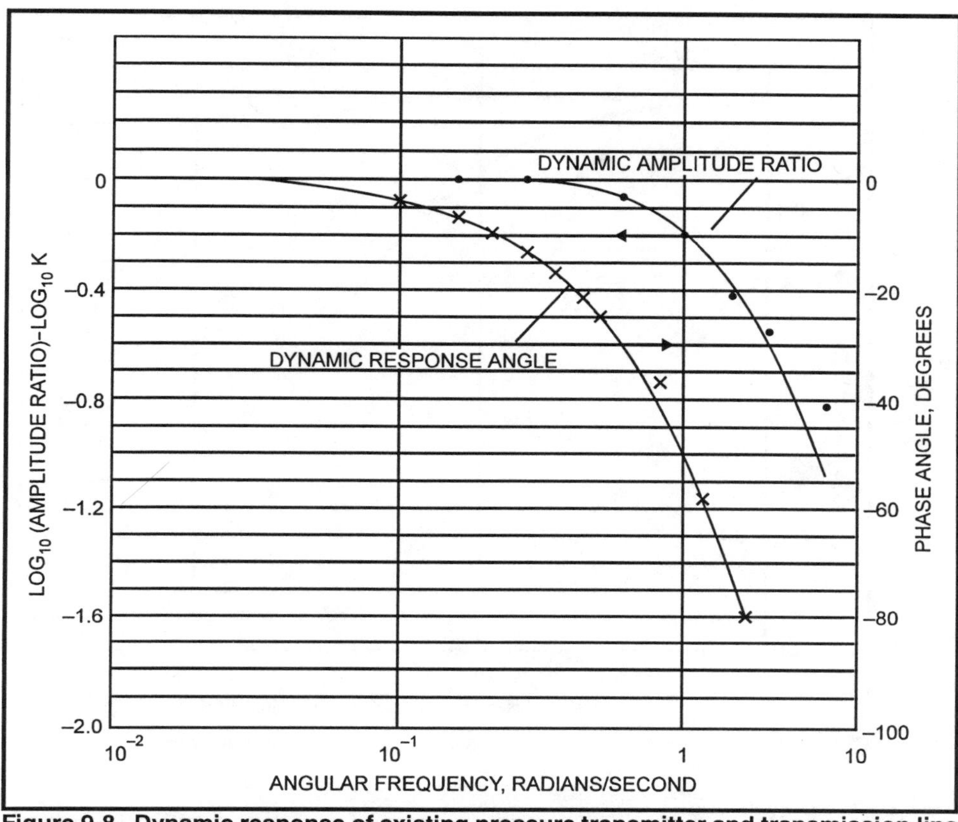

Figure 9-8. Dynamic response of existing pressure transmitter and transmission line

Frequency response information, relating pressures at various points in the system to bleed valve displacement, was also derived from these tests. This was obtained by processing the pulse test data using the Fourier transform computation routine; the results are shown in Table 9-3. The pure delay time in these linearized performance functions is included in the exponential term.

Table 9-3. Computed Frequency Response Functions

$$\frac{\text{Inches water change at premixer}}{\text{\% change in bleed valve displacement}} = \frac{0.19e^{-4s}}{(1 + 5.30s)}$$

$$\frac{\text{Inches water change at Venturi inlet}}{\text{\% change in bleed valve displacement}} = \frac{1.29e^{-s}}{1 + 2.6s}$$

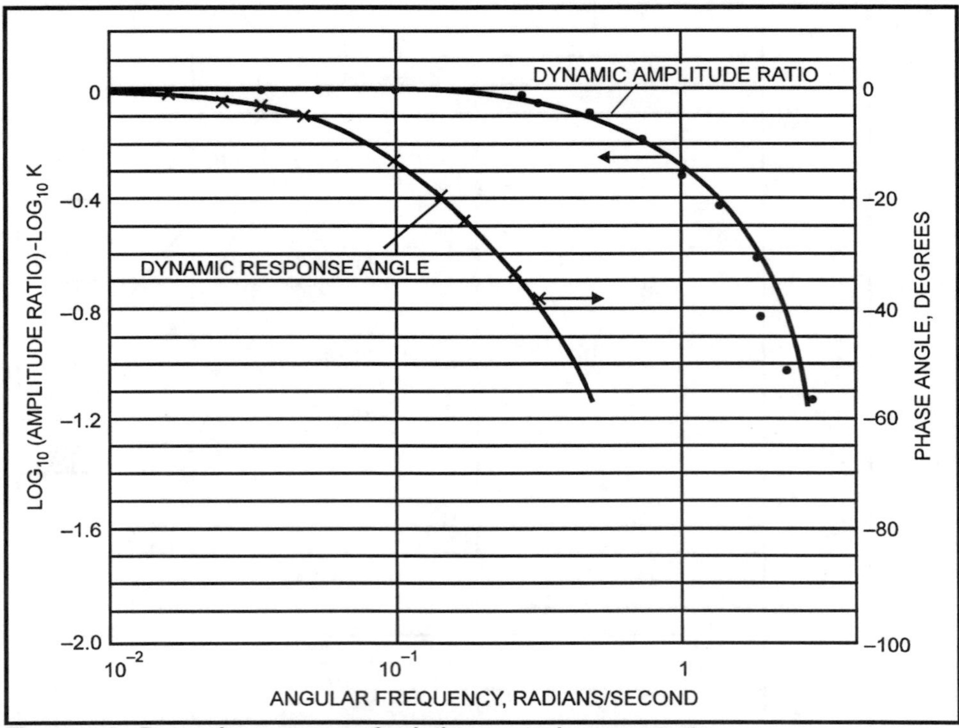

Figure 9-9. Dynamic response of existing control components

It is well-known that the performance of negative feedback control systems deteriorate very quickly as the ratio of delay time to the dominant time constant exceeds 1/10. With a ratio of 0.75 at the premix pressure point, it becomes obvious why the existing control system could not regulate premix pressure by manipulation of the bleed valve.

Design of Pressure Control System

The original control strategy appeared to have merit: delay time and controllers with questionable performance apparently being responsible for the poor performance. The pure delay time of 4 seconds between the bleed valve and the kiln was the principal source of difficulty. Shifting the intermediate pressure sensor closer to the kiln would not only reduce the delay time, but decrease the dominant time constants. A point just below bed 2 was suggested, since this would not only reduce the delay time, but also the volume between the sensing point and the kiln. Even before beginning this investigation, throttling the effluent from the kiln was considered a possibility, providing the motivation for installing the butterfly valve.

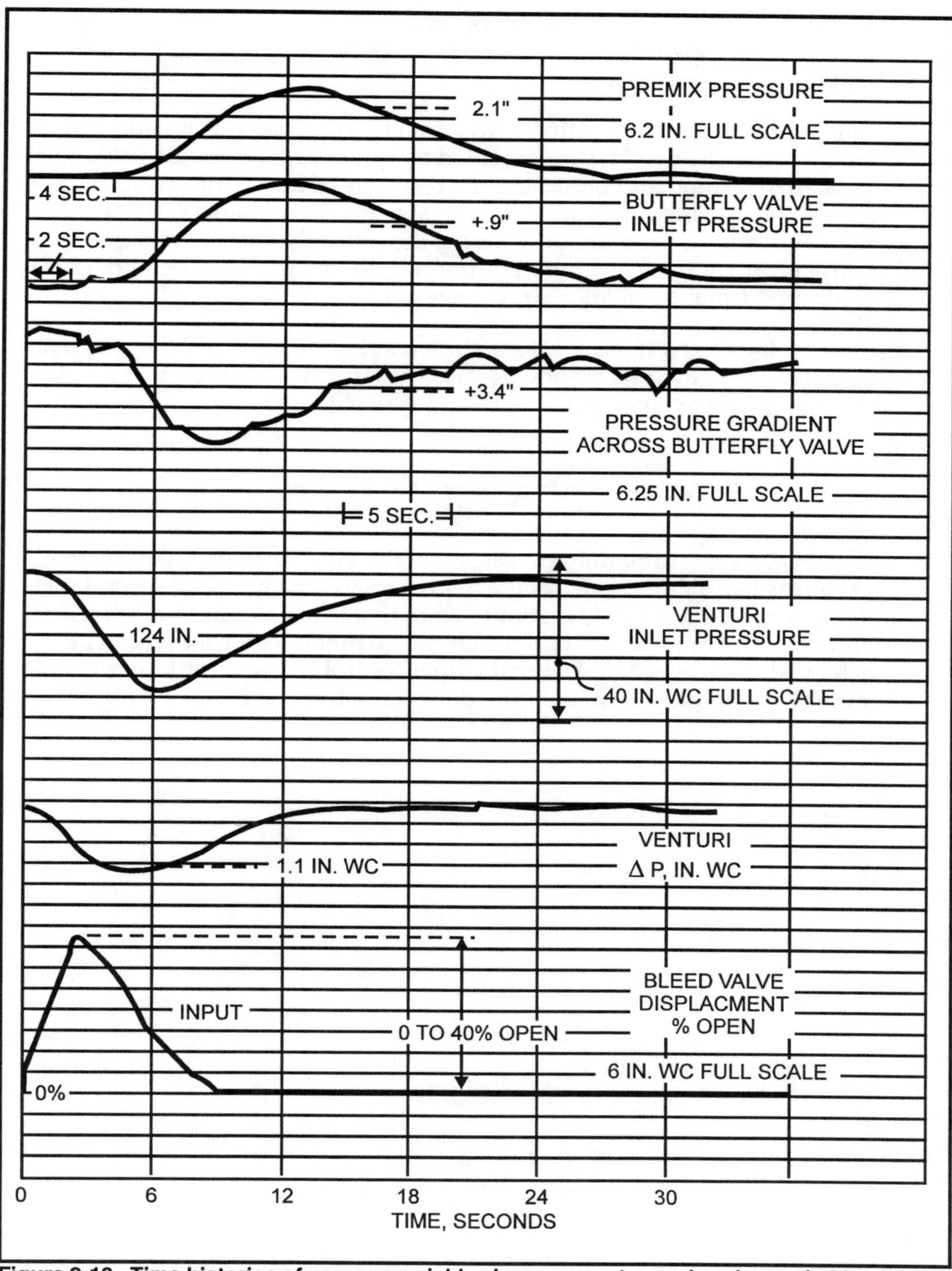

Figure 9-10. Time histories of process variables in response to a pulse change in bleed valve position

For purposes of designing this proposed control system, performance information, especially dynamic response, was required, and towards this end tests were executed.

The butterfly valve was fitted with a transducer so that its position could be measured. The tests were executed by quickly stroking this valve in a pulse-like fashion and observing the response of pertinent variables, notably the premix pressure. Other data were also obtained: of particular interest were the pressure drop across the butterfly valve and the temperature of the gas flowing through it, so that flow rates could be computed from the relation previously presented.

Time histories of a typical test are shown in Figure 9-11, and the frequency response computed from these tests is shown in Figure 9-12, from which the linear approximation, given below, was obtained.

$$PF(s) = \frac{0.027e^{-0.5s}}{(1 + 2.5s)} = \frac{\text{Premix pressure}}{\text{Position of B.V.}}$$

This linear approximation indicates that the kiln-butterfly valve system performs like a first-order system, with a time constant of 2.5 seconds. The delay time of 0.5 seconds is about 1/5 that of the dominant time constant and certainly imposes limitations on the quality of control that can be achieved. However, by placing the final control element for kiln pressure control closer to the kiln, the pure delay time has been reduced 5-fold.

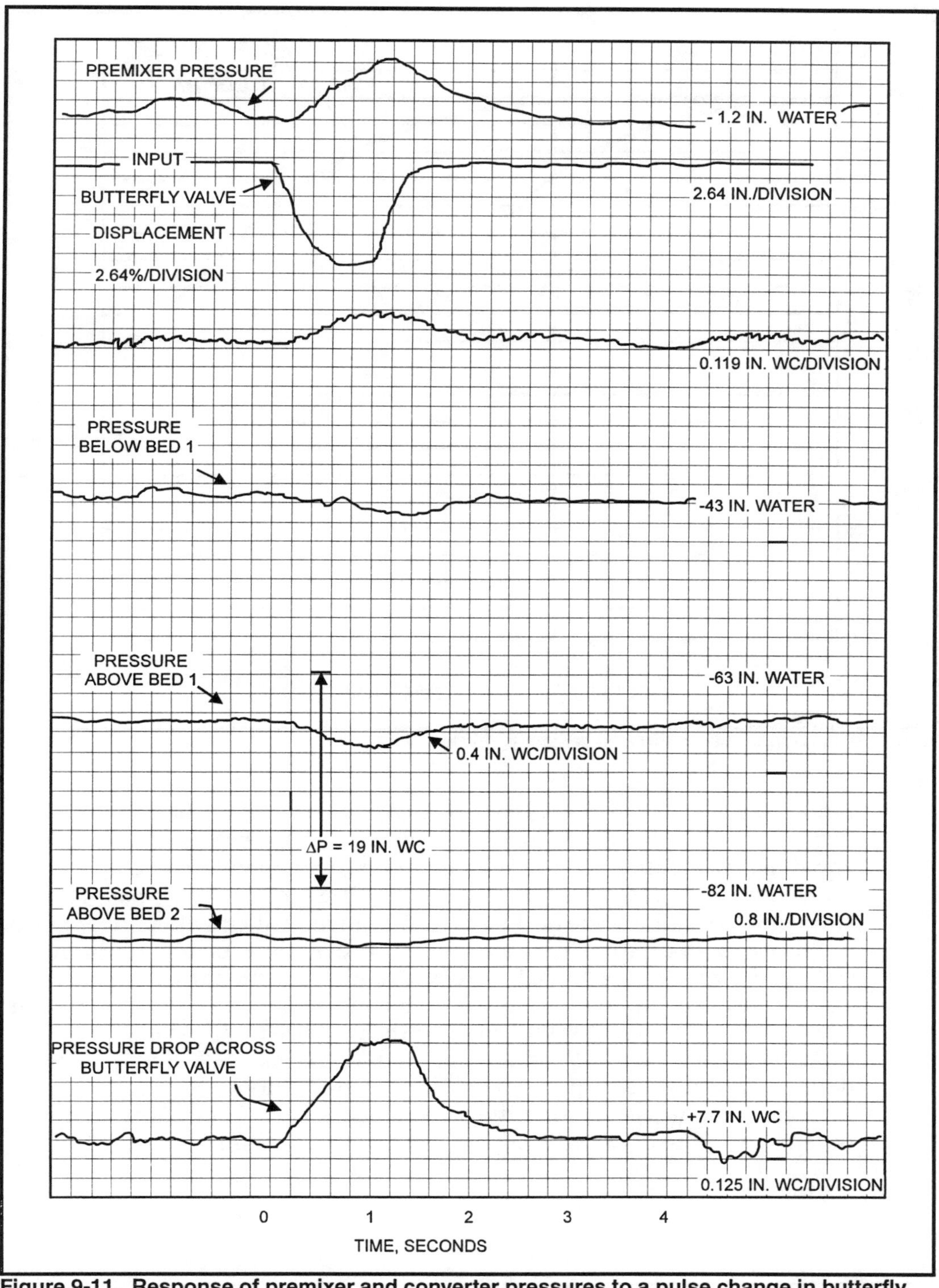

Figure 9-11. Response of premixer and converter pressures to a pulse change in butterfly valve displacement

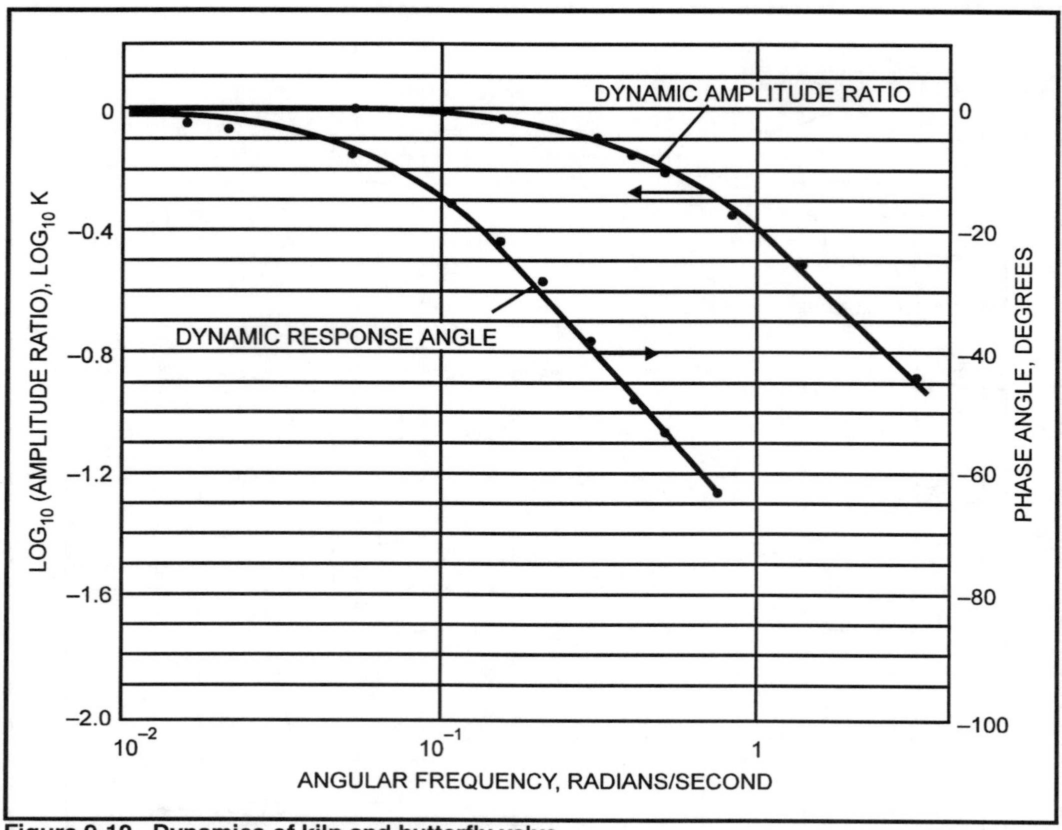

Figure 9-12. Dynamics of kiln and butterfly valve

It may appear that a suitable control algorithm would be the usual P-I form, $K(1 + T_i s)/T_i s$. When this is combined with the process (with T_i set equal to 2.5), an open-loop pure integration function will result, including the delay. The presence of pure delay time may introduce a phase angle sufficient to require some derivative component. Anticipated performance of this closed-loop control system, with no delay time, using optimum settings for both proportional + integral, and proportional + integral + derivative are compared in Table 9-4. The improvement resulting from the use of derivative action is indicated by the increase in the "cross-over" frequency, from 2.7 to 4.9 radians per second. The cross-over frequency is defined here as that frequency at which the closed-loop frequency response becomes unity after the peak of about 2 db. These controller parameters were obtained from the software PCPARSEL[1].

1. PCPARSEL is software, developed by the author, from which controller parameters can be retrieved. See Appendix A.

Table 9-4. Optimum Parameters for Various Controllers for Controlling Kiln Pressure

Controller gain K	Integration parameter T_i	"Derivative" parameter T_d (A=10)	Crossover frequency ω_{co}	Delay time, seconds
3.38	2.5	0	2.7	0
4.4	2.5	0.023	4.9	0

With a P+I controller, best results are obtained if the controller integral parameter, T_i, is set equal to the process time constant, T_1. It has been found, from the relations developed from PCPARSEL, that at any value of DT/T_1, when a value is chosen for T_i, the ratio of $Wc \cdot T_1/K_T$ is approximately 2, where K_T is the total feedforward gain (process plus controller). Therefore, the cross-over frequency $\omega_{co} = 2K_T/T_1$. Also $K_T = 0.676 \cdot T_1/DT$, hence $W_c = 2(0.676) \cdot (T_1/DT) \cdot (1/T_1) = 1.35$ DT. Thus, while the optimum gain depends upon both the time constant T_1, for a first-order system with delay time, or upon the integration factor for a pure integration process with delay time, the cross-over frequency, ω_{co}, depends, inversely, only upon the delay time. The author believes this to be true because when K_T and T_d are optimum, then ω_{co} depends only upon DT.

With no delay time, closing the loop would result in a first-order system, thus permitting a very high controller gain. However, the presence of pure delay time imposes a severe limitation. The three-term algorithm is preferred.

Ideally, the butterfly valve should operate in about the same position over an appreciable range of plant production levels. This is recommended because the gain of the valve would consequently remain fairly constant, particularly if the pressure below bed 2 could be maintained at an appropriate value. To expedite this control strategy, the second (cascaded) control system appeared encouraging. To determine the feasibility of such a scheme, the response of pressures in the converter caused by changes in the position of the bleed valve was needed.

The experimental data (shown in Figure 9-13) were obtained by a pulse change in the position of the bleed valve of about 32%. These time histories are greatly enlarged oscillograph records from which the pure delay time as well as the relative magnitude of each record can be easily observed. Of special note is the magnitude of the change in pressure above bed 1 and the pure delay time of each response.

This information led to the conclusion that the optimum point at which to control the pressure in the reactor was above bed 1. Not only is the response in pressure relatively large (sensitivity high) but the pure delay time is not any greater than that above bed 3. Moreover, there is greater freedom from the high-frequency oscillations that were previously mentioned and were shown as originating below bed 1.

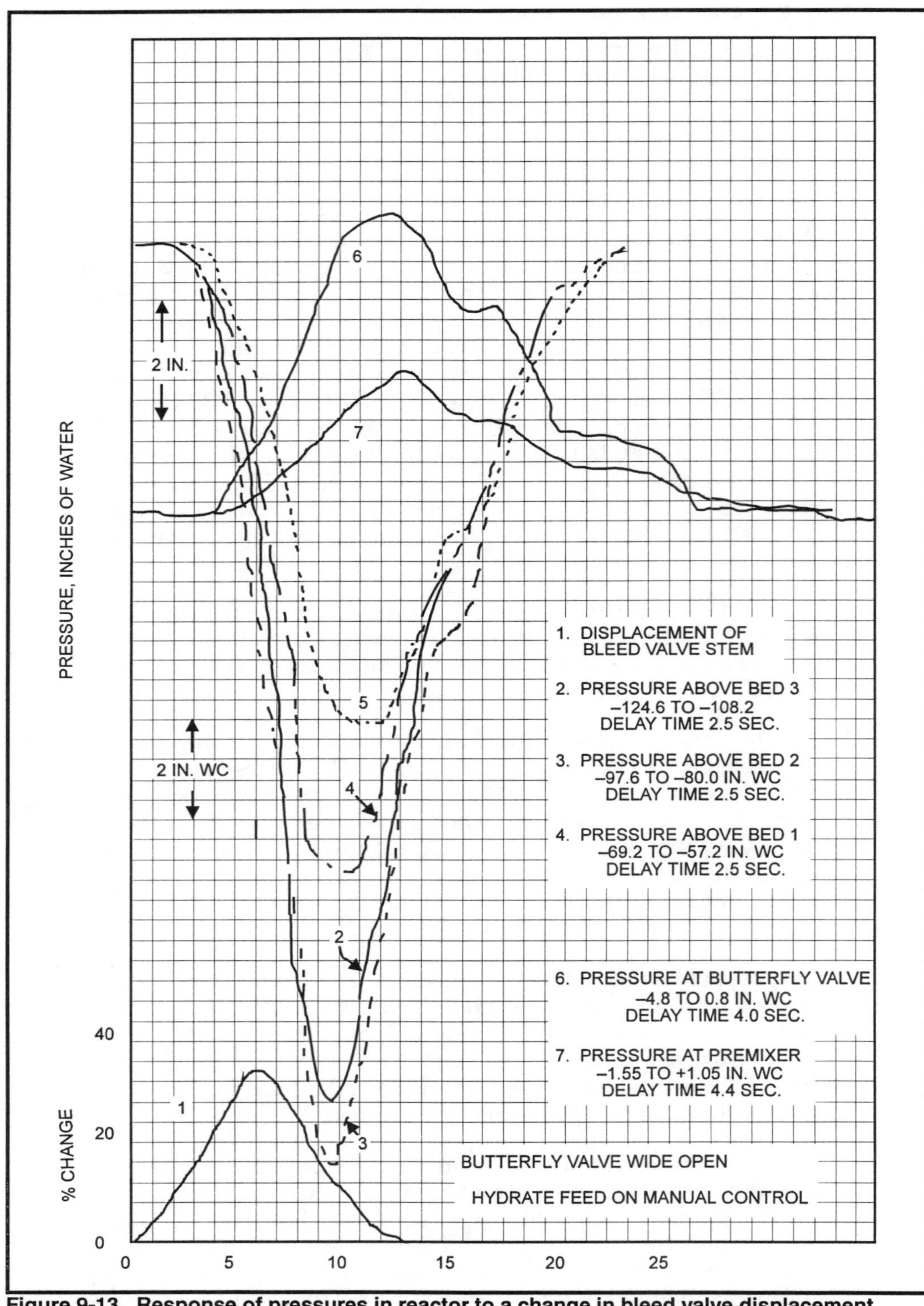

Figure 9-13. Response of pressures in reactor to a change in bleed valve displacement

The frequency response function, derived by processing the time histories of the appropriate records from this test, are shown in Figure 9-14, from which the linear function was derived. Delay time is included in the exponential term

$$PF(s) = \frac{0.555e^{-2.5s}}{(1 + 3.3s)\ (1 + 2s)}$$

Table 9-5 gives the optimum controller parameters for two algorithms assuming delay times of 0, 1, and 2.5 seconds. These parameters were derived from PCPARSEL.

Table 9-5. Optimum Controller Parameters for Various Algorithms and Delay Times

Controller type	Delay time, sec	Process gain	Derivative parameter coefficient	Controller gain	T_i	T_d	Cross-over frequency, ω_{co} rad/sec
P-I	0	0.555	0	5.95	4.7	0	.78
P-I	1	0.555	0	2.50	4.8	0	.50
P-I	2.5	0.555	0	1.63	5.1	0	.33
P-I-D	0	0.555	10	512	3.8	0.05	22.6
P-I-D	1	0.555	10	4.09	3.65	0.19	1.2
P-I-D	2.5	0.555	10	1.64	3.65	0.25	0.65

Of particular interest is the enhancement provided by introducing the derivative component and the deleterious effects of delay time, as shown by the changes in cross-over frequency.

Should the high-frequency oscillations in pressure within the converter in the vicinity of the point at which the pressure sensor is located (in this case the suggested point is above bed 1) introduce "noise," a properly designed filter can be inserted to remove this disturbance before the signal enters the control module.

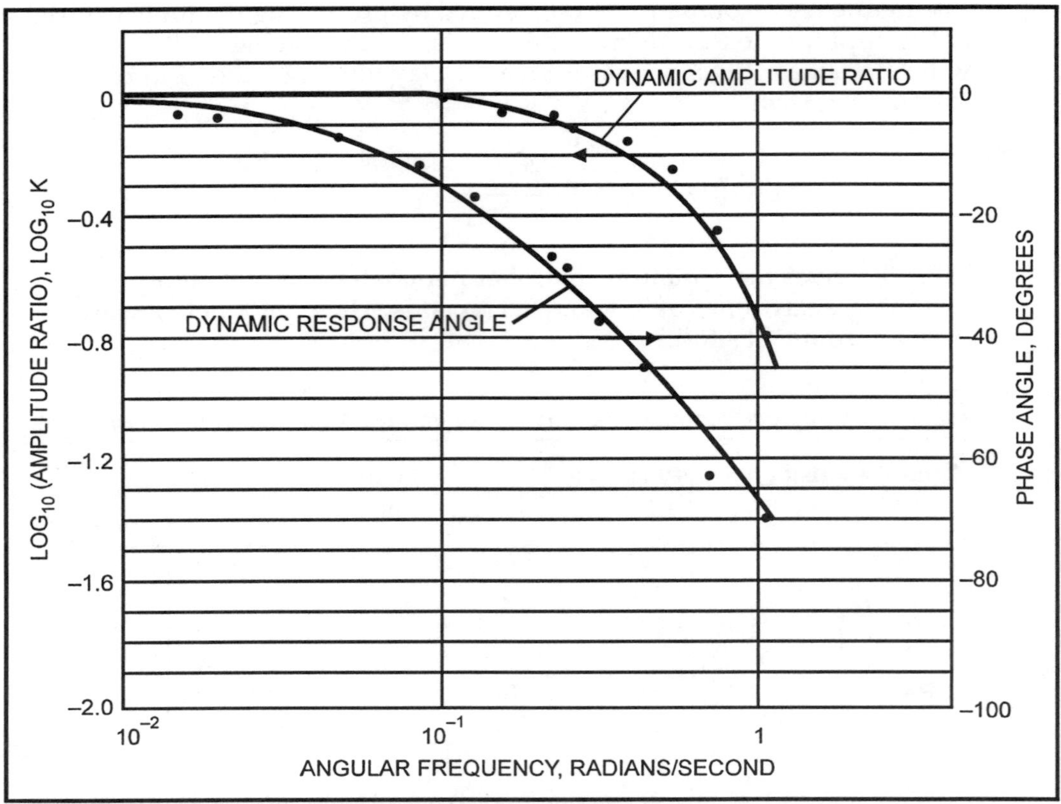

Figure 9-14. Dynamics from bleed valve to above bed no. 1

Summary

- At the time of this study, the procurement of definitive and reliable data from industrial process instrumentation, commonly available, was highly unlikely.
- The test instrumentation installed proved entirely satisfactory, and sufficient information was procured to identify problem areas and to obtain useful steady-state and dynamic information.
- During the testing program, information about the process was obtained that could lead to improvement in its operation and control.
- Data were obtained showing that a small butterfly valve could be calibrated and could then serve as a reliable flow rate sensor. In this instance, the form is especially convenient from a mathematical point of view.

- Data systems with great sensitivity, accuracy, and flexibility are usually required to procure completely satisfactory data from processing plants. This is especially true if dynamic information is desired.
- A revised control strategy for pressure control is presented, along with optimum values for controller parameters for both P-I and P-I-D algorithms, with a number of assumed process delay times.
- In this instance a response from the client was not forthcoming, hence it is not possible to conclude this study with a statement concerning the extent to which this study contributed to improvements in the operation of this process.

10

Ore Preparation in a Kiln Using Multiple Fuels

The preparation for, and execution of, experimental studies of industrial processes sometimes require large investments of both physical and mental energy. The study recounted here was most demanding of both resources. As will become evident, the attainment of the ultimate results was severely limited by conflicting interactions and the marginal performance of some physical apparatus. Perhaps this could have been predicted and much labor eliminated, but then the crucial information might never have been obtained, nor the fundamental conflicts revealed. If the experimental techniques and methodology here described may ultimately be found useful to others, it will be ample reward.

Description of Kiln Operation

Kilns are used in a variety of industries where large quantities of material, usually granular solids, are to be processed at high temperatures for a variety of reasons. The preparation of Portland cement, quick lime, and ore processing are common examples. In each, the objective is to transform the raw material into a form more easily processed in subsequent operations, to promote desired chemical reactions within the material, or to change its physical nature. In preparing Portland cement, for example, the purpose is to heat an intimate and properly proportioned mixture of calcareous and argillaceous material to the point of incipient fusion; the finely pulverized clinker obtained forms the principal component in this product. Perhaps the largest kilns are to be found in the cement industry where they may be 12 to 14 feet in diameter and 200 to 500 feet long.

These kilns are nearly horizontal cylindrical steel vessels, mounted on trunnions, so that they can be slowly rotated. Feed enters the higher end and progresses slowly towards the lower end at a rate roughly proportional to the rate of rotation, which is rarely greater than a few revolutions per minute. Kilns are usually lined with fire brick or other refractory material.

Thermal energy is released within the kiln from fuel combined with air admitted at the lower end. Effluent from the upper end enters the base of a stack, which creates sufficient draft to move the gases through the kiln. Fuel can be oil, natural gas, pulverized coal, other gaseous combustibles, or combinations of these. Air for combustion enters along with the fuel, the flow rate of which is very critical for, in the interest of economy, the excess air should be minimum. Frequently hot air from an independent source is available to supplement that from the surrounding atmosphere.

Since the ends of the kiln cannot be completely sealed from the surroundings, control of pressure within the kiln is used to minimize leakage. Usually this is accomplished by throttling a Venetian-blind-type of valve located between the stack and the end of the kiln. Since these valves may have several vanes, each perhaps 10 to 12 inches wide and several feet long, considerable force is required to actuate the device. Air-operated actuators are commonly used for this purpose. Changes in excess air may be achieved by altering the flow rate of fuel or air, in response to the signal from an oxygen analyzer taking samples of the effluent gases.

The most important information needed to operate a kiln properly is a measure of the quality of the product, the processed solid material. Radiation thermometers, X-ray sensors, direct sampling followed by chemical analysis, or simply visual inspections, of the product are used for this purpose. Very frequently the quality is determined by the judgment of experienced operators, who may merely view the product as it emerges from the kiln.

The kiln considered here was relatively small, being about 150 feet long and 10 feet in diameter. It was used to heat an ore to the point of incipient fusion which, upon crushing, would be suitable for further processing. Three fuels were used: natural gas, pulverized coal, and combustible by-product gas, mostly CO, produced in nearby facilities. The kiln and auxiliaries are shown in Figure 10-1.

The sources of combustion air should be noted: from the surroundings, entering through valve A; from a point in the duct below the rollers, which crush the solid material leaving the kiln; in the effluent of the coal pulverizer. The air accompanying the coal from the mill, in addition to the hot air from the separator, contained fresh air admitted through valve B, which was adjusted manually. Two blowers were involved: blower A for supplying hot air to the burners, bypassing the coal mill, and blower B that supplies the coal mill. Encroachment of air at the firing end of the kiln also occurred because of poorly fitting enclosures, seals, and viewing ports.

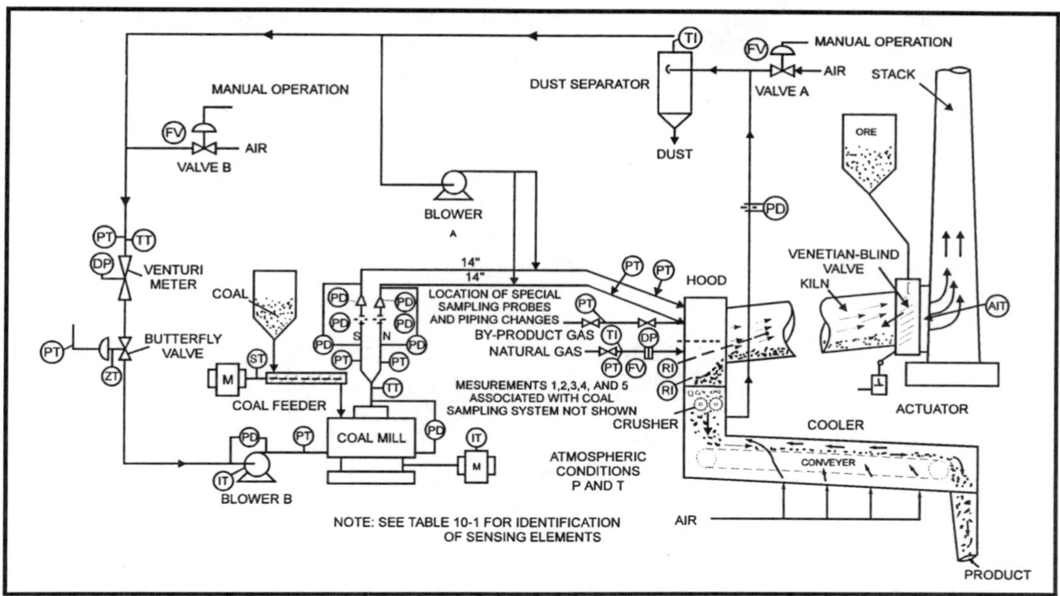

Figure 10-1. Process diagram of original system showing location of test sensors

This cursory description is ample to indicate the sources of many uncontrolled disturbances to the supply rate and condition of both combustion air and fuels. Because existing control systems were not satisfactory, these are not indicated in Figure 10-1. The relative location of test measurements are shown and identified in Table 10-1.

The initial motivation for this study was to evaluate the feasibility of controlling the excess air in the effluent kiln gas by regulating the rate of supply of coal to the coal pulverizer. At this installation the availability of the by-product fuel was subject to frequent changes. This required corresponding changes in both air and other fuel admitted to the kiln The current method for compensating for changes in excess air was to adjust the rate at which coal was fed to the pulverizer. This method was immediately questioned because the response of the coal mill was suspected as being much too slow. However, to demonstrate this, experimental determination of the dynamic response characteristics of the coal feeder and pulverizer system appeared necessary. This obviously would entail not only finding the rate of change of coal from the pulverizer following changes in the rate of coal fed to it from the screw feeder, but also any changes in the size and size distribution of the pulverized coal produced. The latter could possibly influence rates of combustion, as well as temperature distribution, within the kiln.

Description of Mill and Its Operation

The pulverizer was an early model modified Babcock and Wilcox Company mill provided with a centrifugal blower (blower B in

Figure 10-1) for supplying the air, which entrained and transported the coal to the kiln. Within the pulverizer, coal was subdivided by the grinding action of wear-resistant balls about 10 inches in diameter, rotating between two grinding rings. The lower ring, or raceway, was the driven member, power being supplied in this installation by a 100 HP motor, through a pinion gear. The pinion gear meshed with a ring gear to which the lower grinding surface was attached. The upper grinding ring was stationary and was held in position by several equally-spaced heavy springs, which could be adjusted to change the loading on the grinding surfaces.

The pulverized coal was swept out of the mill into the transfer lines by air from the centrifugal blower (blower B in Figure 10-1). The design of a pulverizer is such that air passes around the outer periphery of the ring of balls and transports the crushed coal upward into a size separator located above the grinding section. The separator is a modification of a cyclone, in which the influent is directed tangentially into a calming chamber, so that by virtue of a rotary motion of the fluid stream, together with reduced velocity, the heavy particles settle and re-enter the grinding section. An impeller, attached to the rotating assembly, assists in recycling the heavy material through the grinding section. (Figure 10-2, taken from the manufacturer's literature, illustrates the interior of a mill.)

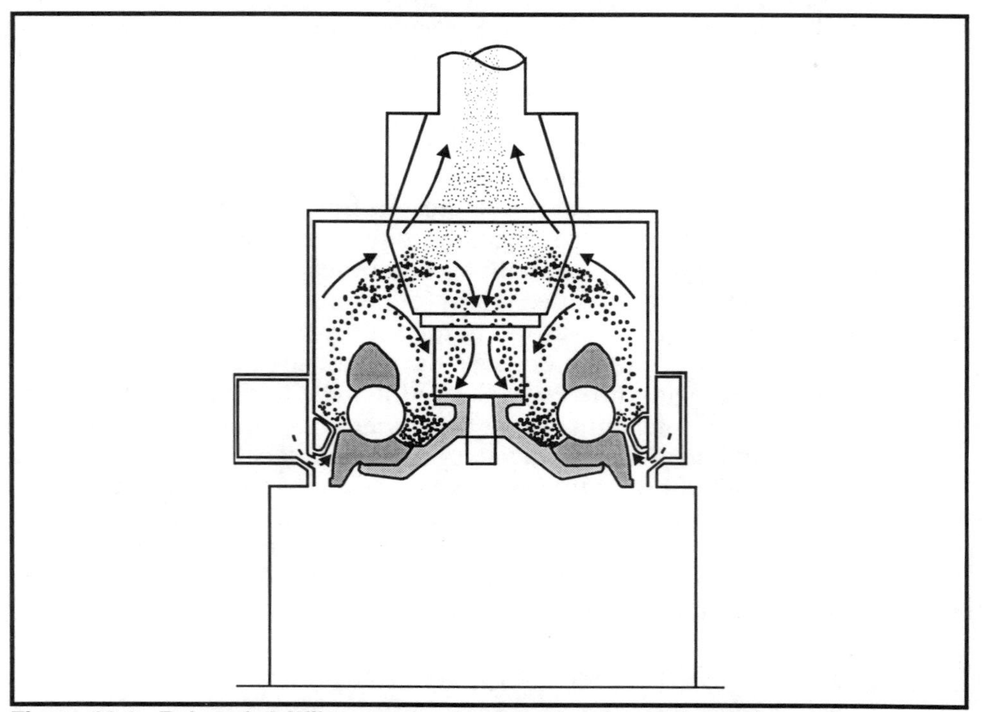

Figure 10-2. Babcock & Wilcox coal pulverizer showing recirculation path

As seen in Figure 10-1, the pulverized coal in the coal mill effluent was conducted to the kiln burners by two parallel transfer lines originating from a "Y" at the top of the mill. These lines, at first vertical, eventually directed the air-coal suspension in a gently sloping manner to the burner section of the kiln.

The air to the mill serves four purposes: (1) removal and transport of the finely divided coal from the mill, (2) classification of solids within the mill in the built-in size separation devices, (3) drying and heating the coal, and (4) providing air at a somewhat elevated temperature for combustion. Four discrete parts of the coal firing system are involved in achieving the desired objectives; each should be designed to be compatible with one another. Thus, the dimensions of the transfer lines should be selected so that the momentum in the air is sufficient to ensure transport of the coal, yet avoid excessive pressure loss. At the same time, the blower must be able to deliver enough air to ensure, at equilibrium, an inventory of coal in the mill that is neither too small nor too great, especially the latter. In addition, the quantity of air passing through the mill must also be such that the elutriating action achieved gives a product of acceptable size and size distribution. Finally the coal feeding system must be capable of regulating the supply of fuel in accordance with demands, but consistent with the capacity, of the pulverizer.

For an existing system, the limits within which the air and coal rates can be varied are rather closely restricted. For example, in order to transport the pulverized coal, a linear velocity in the transfer lines of at least 80 feet per second is required. The flow rate, corresponding to this lower limit in velocity, must be sufficient to prevent excessive accumulation of coal in the mill. If the inventory within the mill is too great, the rate of air delivered by the blower may become limited. For a given flow rate of air there is a maximum coal feed beyond which the mill inventory increases without limit (or until a choking condition develops). At these conditions the size and size distribution approach a limit, tending towards the smallest sizes possible for a given coal. Under conditions corresponding to maximum air flow, there is also an upper limit on the rate at which coal can be added, and particle size is, on average, larger and size distribution greater.

Kiln Performance

Of even greater significance, is the behavior of the kiln in response to changes in quality and feed rate of fuels. While the mill should be capable of supplying the gross amount of fuel required by the kiln, this may not ensure the desired product quality. The size and size distribution of the pulverized coal may influence significantly the quality of product obtained from the kiln. Qualitatively it can be reasoned that the larger the coal particles the longer the zone in which combustion occurs within the kiln, and the smaller the particles the more rapid the combustion, and the zone of highest flame temperature will be closer to the firing end of the kiln. Temperature and temperature distributions within the kiln obviously determine the physical and chemical reactions occurring therein and, ultimately, the nature of the product.

Thus, it is apparent that establishing and maintaining optimum conditions within the kiln becomes a very difficult task, which requires careful regulation of fuel rates, mill operation, excess air, and rate of kiln rotation.

Tests, Instrumentation, and Analytical Procedures

The tests conducted on the system and components naturally divide into two parts, static and dynamic. The former describe the steady state, or time invariant behavior, of a system at various operating conditions; the latter focuses on transient performance, that is, time-dependent properties—how long it takes to effect changes, or what the responses are, time-wise, to changes in independent variables. The details of the experimental procedures and results procured from these tests are presented below.

Table 10-1 identifies the sensors; Figure 10-1 shows their location. All test transducer outputs were recorded on multi-channel oscillographs equipped with high-gain, high-impedance dc amplifiers, appropriate filtering, amplification, suppression, and chart speed adjustments. Transducers were calibrated either with laboratory or plant calibration devices. Pressure and pressure differentials were measured with strain gage sensing elements; temperatures with bare, ungrounded, iron-constantan, Type-J thermocouples. Two types of radiation pyrometers were used for procuring indications of flame and solid product temperatures. The data recorders were located in a small building somewhat removed from the processing apparatus. The common electrical ground was located near the data system.

The most difficult measuring task was that of determining the rate at which coal was delivered to the kiln. The direct calibration of the existing coal feeder was not considered feasible, hence a technique was developed for procuring representative samples from the coal transfer lines, indicative of the mass flow rate through them.

When sampling finely divided solids suspended in moving gas streams, withdrawal must be made without disturbing the flow pattern in the vicinity of the sampling point. This requirement is satisfied if the velocity of withdrawal is identical to the velocity of the fluid stream in the vicinity of the point of sampling. The condition is termed isokinetic. If isokinetic conditions do not exist, and the velocity in the probe is too low, an excess of solids with a greater proportion of material heavier than the average may be captured. This occurs because, as the gas stream is partly diverted around the probe, the larger sizes resist deflection and enter the probe in higher proportion than in the stream at large. The opposite effect occurs if the sampling velocity exceeds that of the surrounding environment.

Table 10-1. Measurements and Sensors

Identification	Measured variable	Sensing element
PT	Sample drum pressure	Statham ±25 psi
PT	Sample probe ΔP, North	Statham ±1 psi
PT	Sample probe ΔP, South	Statham ±1 psi
DP	Sample flow rate	Statham ±2.5 psi
TT	Sample temperature	Aeroresearch TC
PD	Orifice ΔP, N	Statham ±5 psi
PD	Orifice plus nozzle, ΔP, N	Statham ±5 psi
PD	Orifice ΔP, S	Statham ±5 psi
PD	Orifice plus nozzle, ΔP, S	Statham ±5 psi
DP	N Line ΔP, Mill to sample	Statham ± 2.5 psi
DP	S Line ΔP, Mill to sample	Statham ± 2.5 psi
PT	Pressure, before nozzle, N	Statham 0-100 psia
PT	Pressure, before nozzle, S	Statham 0-100 psia
PT	Pressure at burner, N	Statham 0-100 psia
PT	Pressure at burner, S	Statham 0-100 psia
PI	Atmospheric pressure	Statham 0-100 psia
PI	Atmospheric temperature	Aeroresearch TC
PT	Flame hood pressure	Statham ± 1 psi
RI	Flame temperature	Radiation sensor
RI	Product temperature	Radiation sensor
TI	Separator temperature	Aeroresearch TC
PT	Bypass air orifice ΔP	Statham ± 5 psi
TT	Venturi inlet temperature	Aeroresearch TC
PT	Venturi inlet pressure	Statham ± 1 psi
DP	Venturi ΔP	Statham ± 5 psi
PT	Valve diaphragm pressure	Statham ± 25 psi
ZT	Valve stem displacement	Helipot
PT	Air blower ΔP	Statham ± 25 psi
IT	Blower motor current	Recording ammeter
PT	Mill air inlet pressure	Statham ± 25 psi
DP	Mill ΔP	Statham ± 25 psi
TT	Mill outlet temperature	Aeroresearch TC
TT	Mill outlet temperature	Aeroresearch TC
ST	Coal screw RPM	Tachometer
IT	Mill motor current	Recording ammeter
PT	Byproduct gas pressure	Statham ± 25 psi
TI	Byproduct temperature	Aeroresearch TC
DP	Byproduct Venturi ΔP	Statham ± 5 psi
TI	Natural gas temperature	Aeroresearch TC
PT	Natural gas pressure	Statham ± 25 psi
DP	Natural gas flow orifice ΔP	Statham ± 5 psi
AIT^{02}	Oxygen in stack gas	Oxygen analyzer

While the probe must be as small as possible, and sharp-edged, excessive time must not be required to obtain samples large enough for subsequent analyses.

Fortunately this problem has received extensive attention by others so that considerable excellent literature is available. Two sources were particularly helpful.[1]

Two probes were constructed and proved to be completely satisfactory. Probe configurations are shown in Figure 10-3. This figure also shows how these probes were attached to, and positioned in, the transfer lines. Other parts of the sampling system must also operate satisfactorily, requiring careful consideration in their design and construction. The arrangement is shown here in some detail because of the possible interest to others who may have need for such sampling systems.

Sampling of the coal from the transfer lines was not only needed for size and size distribution analyses, but for determining the flow rate of coal through the transfer lines. Samples were desired at a number of steady-state conditions, as well as following abrupt changes in the rates of air and coal to the mill. From these latter samples, taken in rapid succession, the transient response of the mill was determined. For this latter purpose, samples had to be taken from positions in the transfer lines that were representative of the average concentration across the entire cross-section. These rather stringent requirements led to the system finally developed and which is illustrated in Figure 10-4.

---

1. "Pulverized Coal Transport through Pipes," Patterson, R. C., Combustion, July, 1958, p. 47.

   "Isokinetic Sampling Probes," Dennis, Richard, William R. Samples, David M. Anderson and Leslie Silverman. Ind. Eng. Chem. 49, No. 2, Feb. 1957, pp. 294-302.

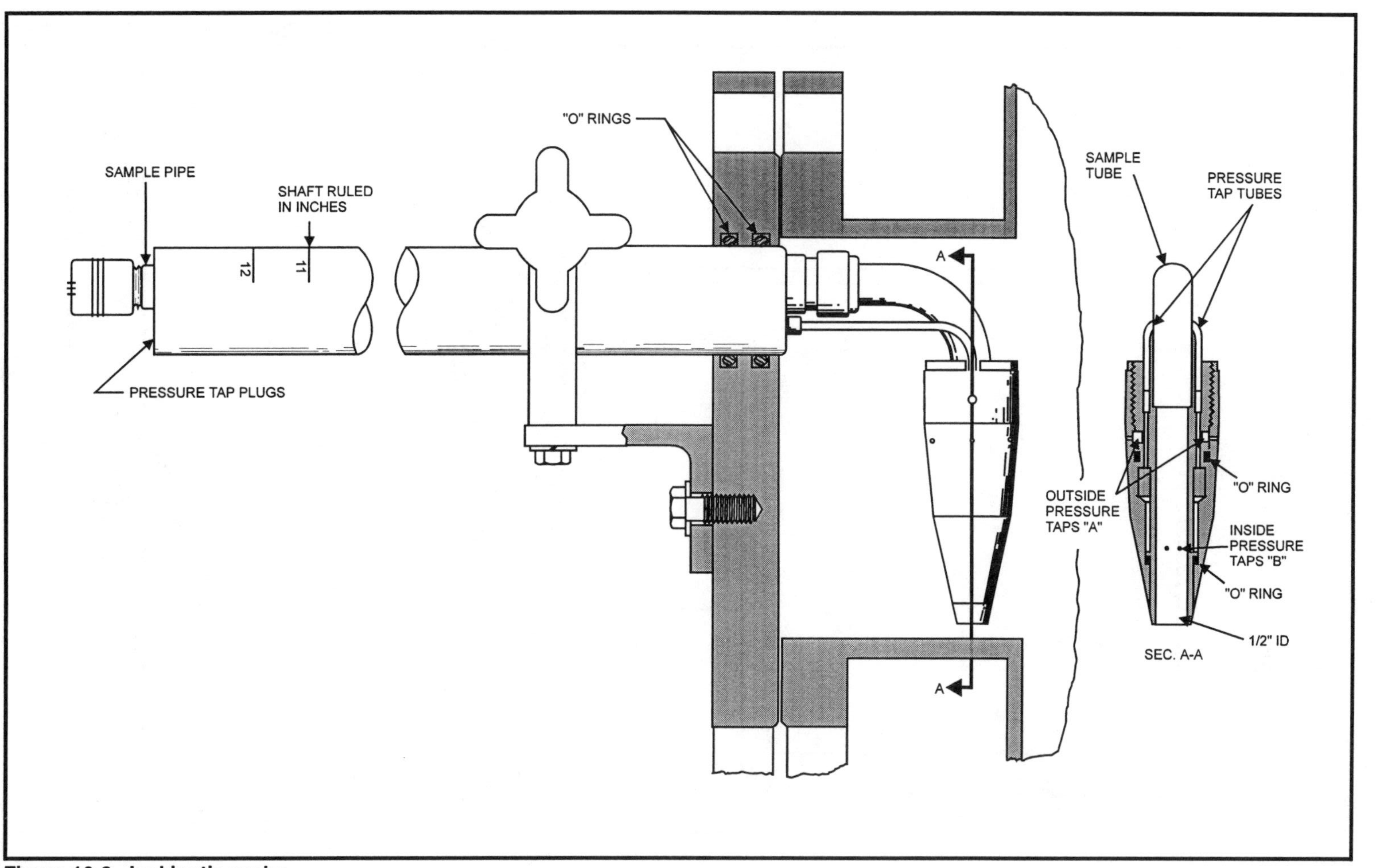

Figure 10-3. Isokinetic probe

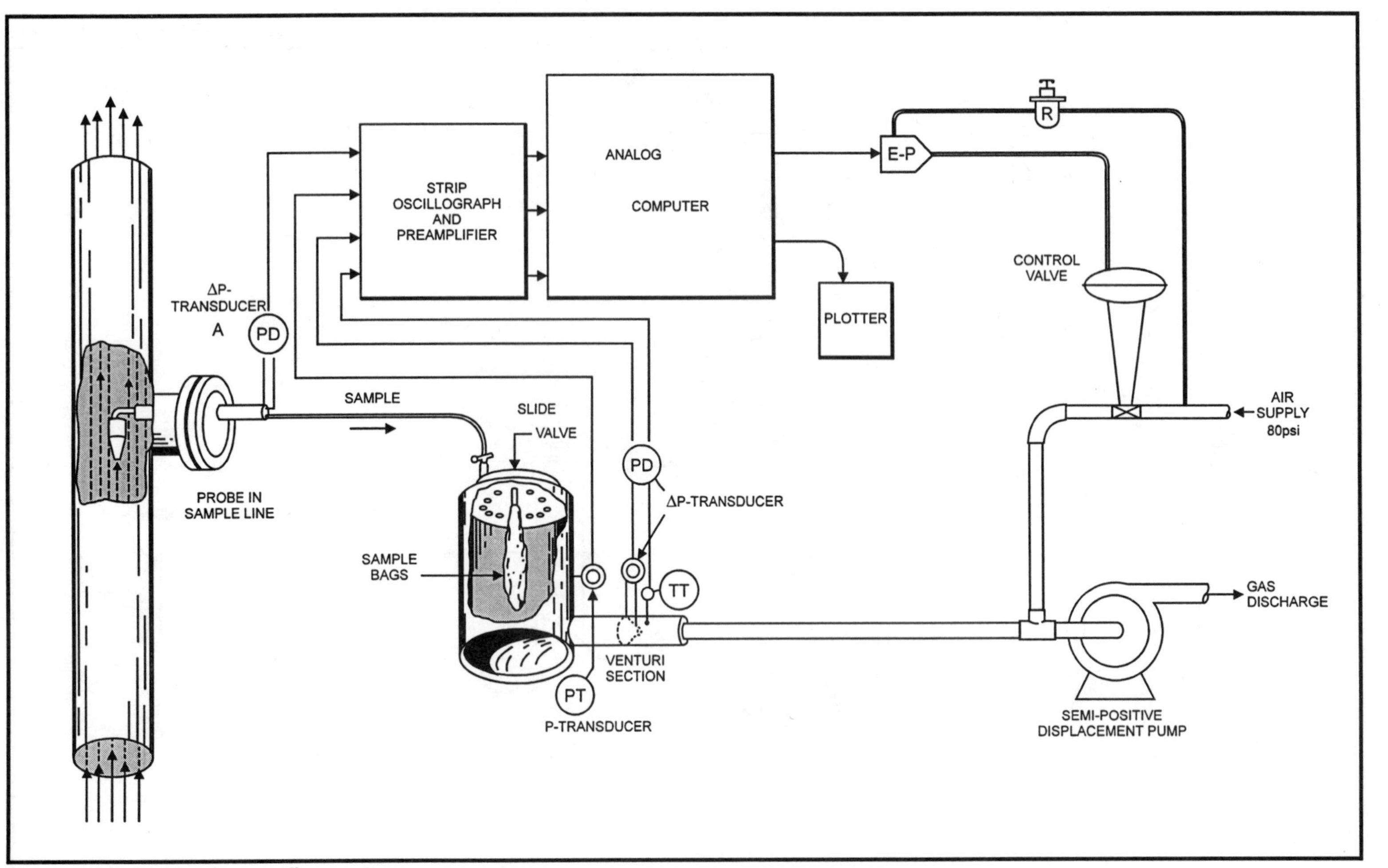

Figure 10-4. Isokinetic sampling system

Construction of Isokinetic Coal Sampling System

On the top of a 55-gallon drum was attached a flat circular plate in which were 12 equally-spaced, threaded holes with center-lines on a 15-inch circle; 3/4-inch pipe nipples were inserted into these holes extending into the drum. Filter bags for collecting samples of the solid material could be secured to these extensions. Another identical circular plate, with a single hole, was placed over the previous plate, the latter secured to the lower plate at the center by a pin. A flexible conduit (1/2-inch polyethylene tubing) leading from the sampling probe was attached to a pipe nipple extending from the top of the upper plate. By rotating the top plate the sampling conduit could be quickly aligned with a hole in the lower plate, directing the flow into a specific receiving bag.

Isokinetic conditions require that the pressure at point A be equal to that at point B (see Figure 10-3). As the velocity in the vicinity of the probe changes, similar changes must be made in the velocity within the probe so that the difference in pressure between A and B remains zero. To achieve this manually appeared unrealistic in view of the rapid fluctuations in local velocity that normally exist. Consequently, control of this differential pressure was made automatic. This was accomplished by using the signal from the probe differential pressure transducer to regulate the position of the motor valve admitting 80 psi air to the suction of the vacuum pump. A small Nash Hy-Tor™ compressor (driven at 3600 rpm by a 5 HP, 3-phase, 220v motor) created the pressure differential by which the sample was drawn into the probe. By regulating the rate of air entering the suction of this pump, from the 80 psi source, the pressure gradient in the probe, which signified isokinetic conditions, could be controlled and maintained even during brief periods of sampling.

The probe differential pressure, along with the pressure in the barrel, the pressure drop across a Venturi meter in the suction line, and the temperature of the air at that point were all recorded on individual channels of the oscillograph. These measurements were made so that the quantity of air associated with each sample could be computed. Sensor signals were also directed to a portable analog computer, from which an appropriate signal was produced as input to an electronic-to-pneumatic relay. This relay regulated the pressure to the diaphragm of the control valve admitting air to the semi-positive displacement pump. (Several other signals were also recorded as noted in Figure 10-4.)

This system proved so responsive in following velocity fluctuations at the sample probe that the pressure drop across the Venturi nozzle, used to measure the rate of sampling, was subject to oscillations of considerable magnitude. Thus, direct reading of the flow rate signal would have been difficult and uncertain. Accordingly, this signal was processed in another section of the analog computer so that the value of $\int_0^t \frac{\sqrt{\Delta P_s}}{t} dt$ was produced. This was continuously presented on an X-Y recorder as a function of sample time, *t*. Thus, at the end of a sample period, the average value of $(\sqrt{\Delta P})_{ave}$ could be found by reading the terminal value on the "Y" coordinate. The control and computing circuits are shown in Figures 10-5 and 10-6. A typical record, obtained during an actual test, is presented in Figure 10-7. This record shows that about 5 seconds were required to establish isokinetic conditions, and that the probe differential pressure was controlled within less than 1/2 in. WC during the sampling interval. The system performed satisfactorily throughout the entire test program.

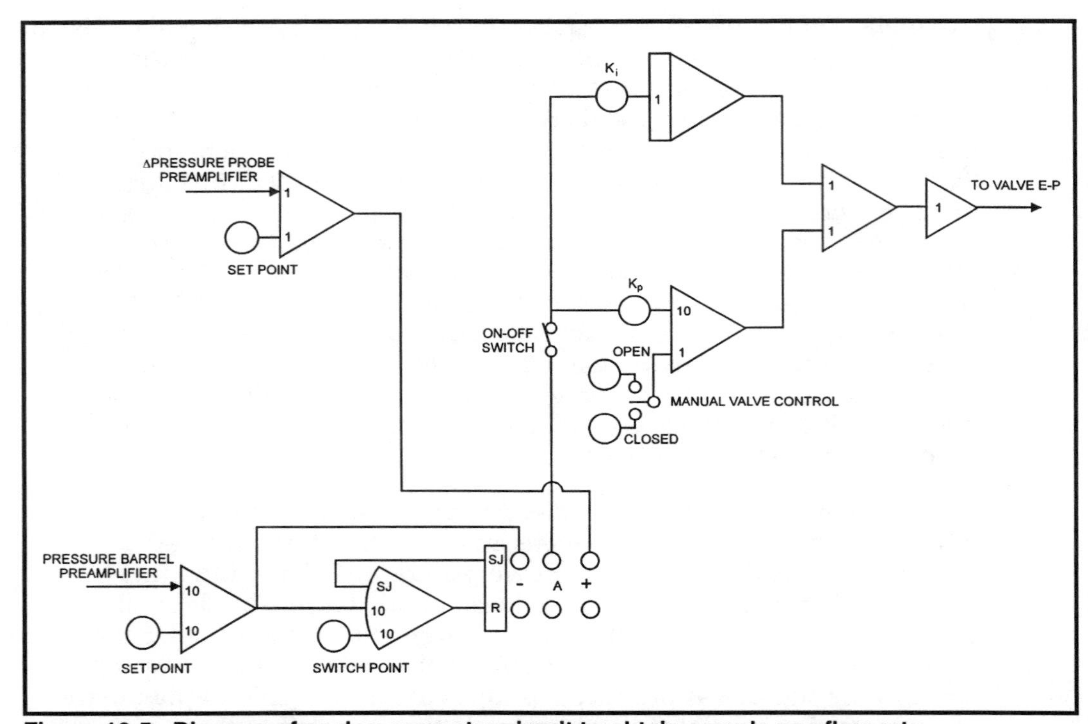

Figure 10-5. Diagram of analog computer circuit to obtain sample gas flow rate

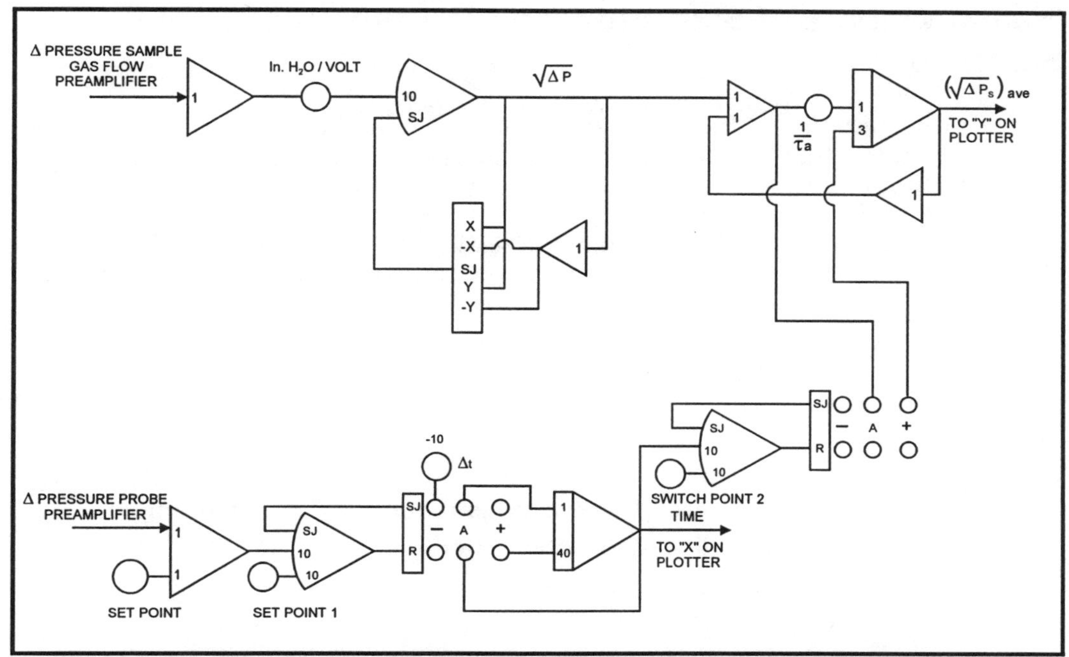

Figure 10-6. Diagram of analog computer circuit to obtain sample gas flow rate

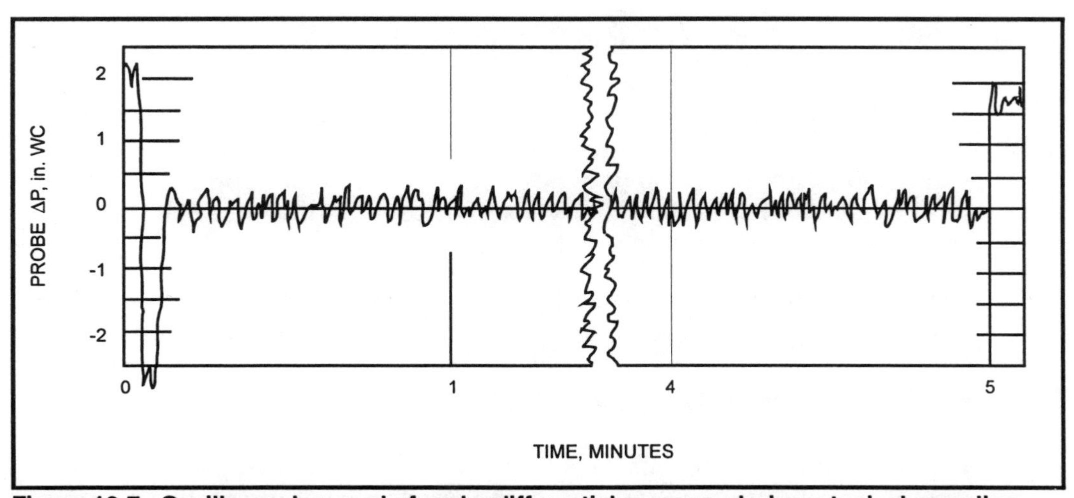

Figure 10-7. Oscillograph record of probe differential pressure during a typical sampling period

Measurements of Size and Size Distribution of Coal Samples

Determination of the size and size distribution of the pulverized coal was another difficult problem requiring special expertise. Sedimentation techniques for size analysis were chosen for most of this work.[2]

For gravimetric sedimentation, the relation between settling time and the diameter of an equivalent sphere in microns is

$$t = \frac{1.8 n 10^7}{(\rho_1 - \rho_2) D^2}$$

where

n	=	viscosity of sedimentation medium, centipoises
ρ_1	=	density of solid material
ρ_2	=	density of medium
D	=	diameter of equivalent spheres, microns
t	=	settling time, seconds

For centrifugal sedimentation

$$t = C + \frac{2.30 \times 10^{11} \eta}{3.6 \times 10^5 (\rho_1 - \rho_2) D^2}$$

where C is an experimentally determined constant associated with the acceleration and deceleration rates of the particular centrifuge used. The coefficient, 2.3×10^{11}, also depends upon the particular apparatus employed. A reference angular velocity is used, in this case 600 rpm, which accounts for the factor 3.6×10^5.

Particle density was determined by a standard air permeability technique and was found to be 1.665 grams/cc for the samples submitted.

Performance of Existing Process

Records of the behavior of the system were first obtained under normal operating conditions. Almost all of the recorded signals exhibited random variations of considerable magnitude, indicating poor control, or complete lack thereof. Upon placing the controllers in the manual mode, the

---

2. These techniques were described in:
Particle Size: Measurement, Interpretation and Application, Riyad R. Irani and Clayton F. Callis, John Wiley & Sons, 1963.
Micromeritics, J. M. Dallavalle, Pitman Publishing Corp. (1948)

variations immediately ceased; constant conditions would prevail until a change, intentional or not, occurred somewhere within the system. In the interest of space these records are not shown.

This state of poor performance and instability was caused by one or more of the following.

- Measuring elements that were improperly located.
- Measuring elements that did not respond rapidly enough.
- Inferior dynamic characteristics of valve actuators.
- Pure delay time between various inputs and outputs.
- Ill-conceived control strategies and improper arrangements of control components.
- Non-linear sensitivities of control components, especially control valves.
- Insufficient capacity of some processing equipment under some conditions.
- Improper adjustments of controllers and/or choice of controller parameters.

Deficiencies in the existing control systems were amply demonstrated by the oscillograph records previously mentioned. Unregulated feedback paths from the kiln to the coal firing system probably occurred. For example, if a change in coal firing was demanded, by a signal from the oxygen analyzer taking samples from the exit end of the kiln, this would also change the flow rate of primary air through the coal mill because of the change in resistance to air flowing through it. This, in turn, would change the amount of excess air and the oxygen content in combustion gases and, thus, change the rate of coal fired, perpetuating the cycle. Moreover, the oxygen content in the kiln effluent could vary rapidly in response to any of a number of changes. These changes could occur at several points in the system, any of which might occur at a rate exceeding the analyzer's ability to respond. In addition, the significant time lag in detecting variations in oxygen by the existing analyzer rendered it inadequate for control purposes.

The interdependence of the firing system and the kiln system should be reduced. Ideally, the coal mill should be supplied with an independent source of preheated air at controlled temperature and flow rate. This, and a number of other quite obvious changes, would make for a more controllable arrangement and eliminate interaction between and among subsystems. The need for energy conservation was, however, recognized.

Before quantitative conclusions could be made about a revised control scheme, the performance of the coal mill had to be determined. Especially important was the capability of the mill to respond to changes in the rate of coal fed to it. Description of this work occupies a major part of the remainder of this presentation.

Improving Sampling Environment in Transfer Lines

Early in this study the discovery was made that the concentration of coal varied as much as 50-fold across the diameter of the transfer lines. In addition, the distribution of coal between the two lines was not uniform. Both of these situations, especially the former, made it impossible to obtain reliable information on the rate of coal transported by means of the previously described isokinetic sampling system. Accordingly, the piping was rearranged, and a re-entrainment section was installed in each transfer line. These re-entrainment sections consisted of an orifice plate with a 9-inch orifice followed by a nozzle with a 10-inch throat, illustrated in Figure 10-8. The first created mixing, the latter not only increased the velocity, but ensured a more uniform velocity distribution in the transfer lines. As a result, satisfactory samples from any point across the diameter of the nozzle could be procured. In addition, distribution of coal between the two transfer lines was equalized.

Calibration of Butterfly Valve for Metering Air Flow to Coal Mill

Preliminary test data showed that the Venturi meter used to measure the flow rate of air to the coal mill was unreliable. Flow rates computed using measurements from this meter were both inconsistent and confusing. At best it appeared that the discharge coefficient did not remain constant. Fortunately an alternative method of metering this air stream was found.

As part of the testing program, ambient air was forced through the coal mill, with no coal being added. Among the measurements made during these tests were the pressure at, and pressure gradients across, the orifice plates in the mixing sections. At the same time the pressure at, and the pressure gradient across, the butterfly valve in the air supply line were measured, along with measurements of the angular position of its vane. Assuming the orifice plates to have discharge coefficients of 0.61, the total air flow rate was computed at several conditions. Using flow rates obtained in this manner the 21-inch butterfly valve was calibrated. The results are shown in Figure 10-9, from which the relation below was derived.

$$Q_v = 178e^{0.043z_v}\sqrt{\Delta P_v/\rho_v}$$

where

Q_v = cu. ft./min. at flowing conditions

z_v = valve stem position, % open

ΔP_v = pressure gradient across valve

ρ_v = density at valve, lb/cu.ft.

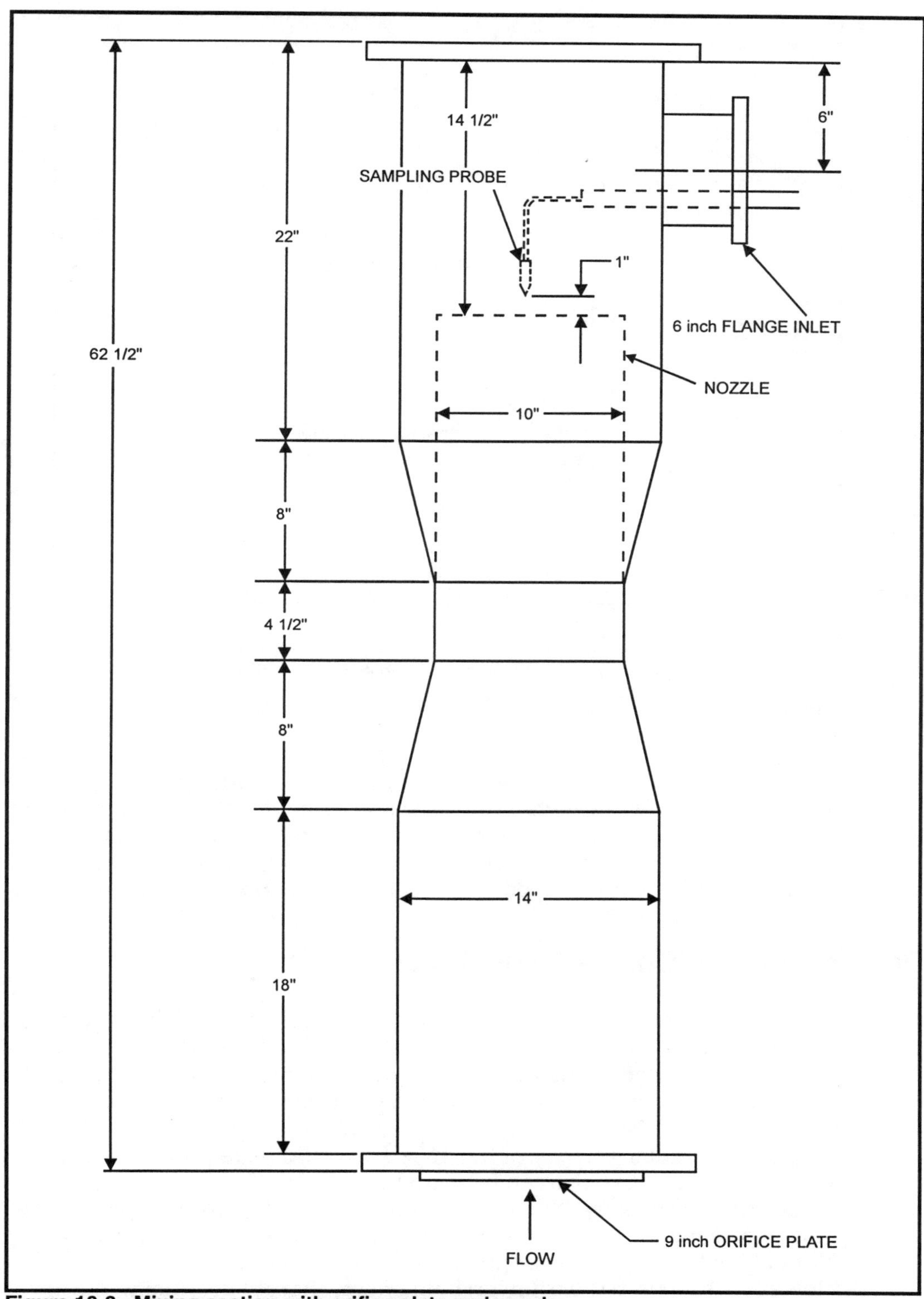

Figure 10-8. Mixing section with orifice plate and nozzle

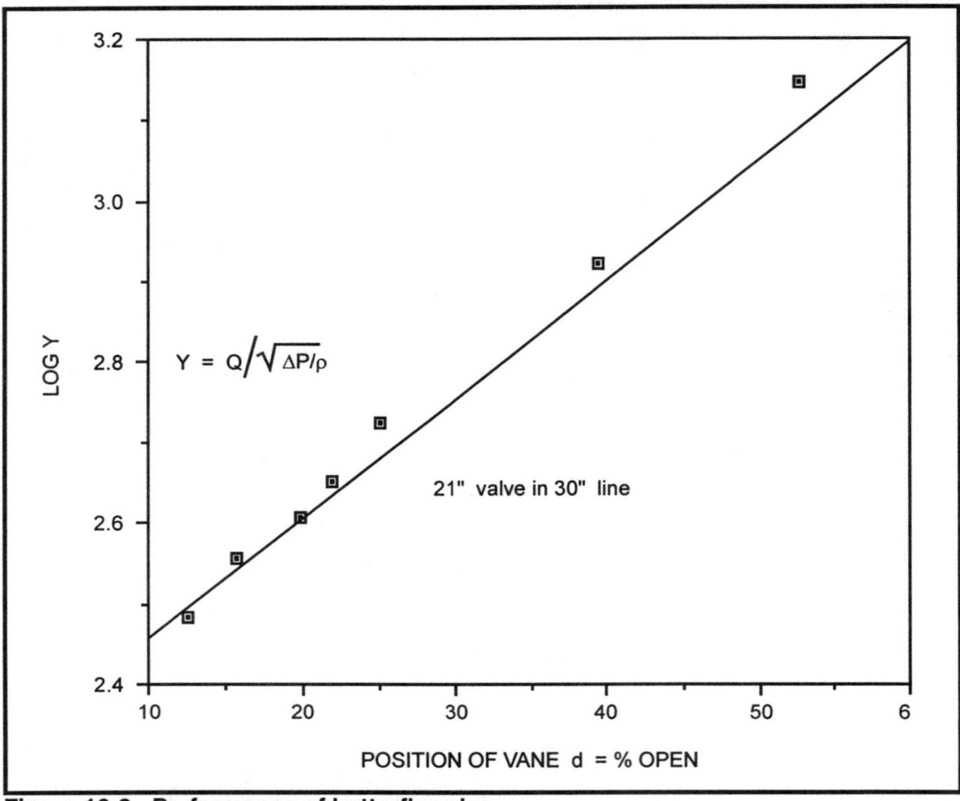

Figure 10-9. Performance of butterfly valve

In subsequent tests where coal was fired, and these orifice plates could not be used as air flow rate meters, the flow rate of air was computed from this relationship. The probable cause of the inconsistencies in the flow rates computed from the Venturi meter data was its poor location, which was about 40 inches upstream from a 90-degree elbow and only 36 inches from the butterfly valve.

Performance of Coal Mill Air Blower

Air was delivered to the coal mill by a centrifugal blower, driven by a 125 volt, 440 volt motor, rated at 1750 RPM. Figure 10-10 shows experimentally determined volumetric performance of this blower for both cool and hot air. The diminution in performance caused by an increase in temperature at the intake is readily apparent. A substantial increase in angular velocity of the impeller would be required to deliver 12,000 cfm of hot gas, with an increase in pressure of 60 in. WC. Figure 10-11 also shows the severe diminution in the performance of this blower as a mass delivery device, which occurs when handling hot air. Not only is the volume of air greater at higher temperatures, but the blower suction pressure is lower; both effects sharply decreasing the capacity of this unit.

Figure 10-12 completes the description of the blower performance and, again, shows the effect of increased inlet temperature. In this figure the rate at which energy is imparted to the air to increase its pressure is plotted against a measure of the current to the driving motor. When moving cool air, with densities ranging from 0.0631 to 0.0693, the current signal reaches a maximum of 32; whereas with hot air, at densities from 0.040 to 0.046, the maximum current function is about 24, a reduction of 25% in this signal. In terms of the mass flow rate, this corresponds to a reduction of 35%, from 810 to 530 lbs of air per minute.

Increased mass flow rate could be achieved by significantly increasing the impeller speed, replacing the impeller with a more efficient design, or moving the throttle valve from the suction to the discharge. Lowering the temperature of the air to the suction appeared to be the most practical solution for this installation, representing a trade-off between increased fuel and sensible heat recovered in hot air.

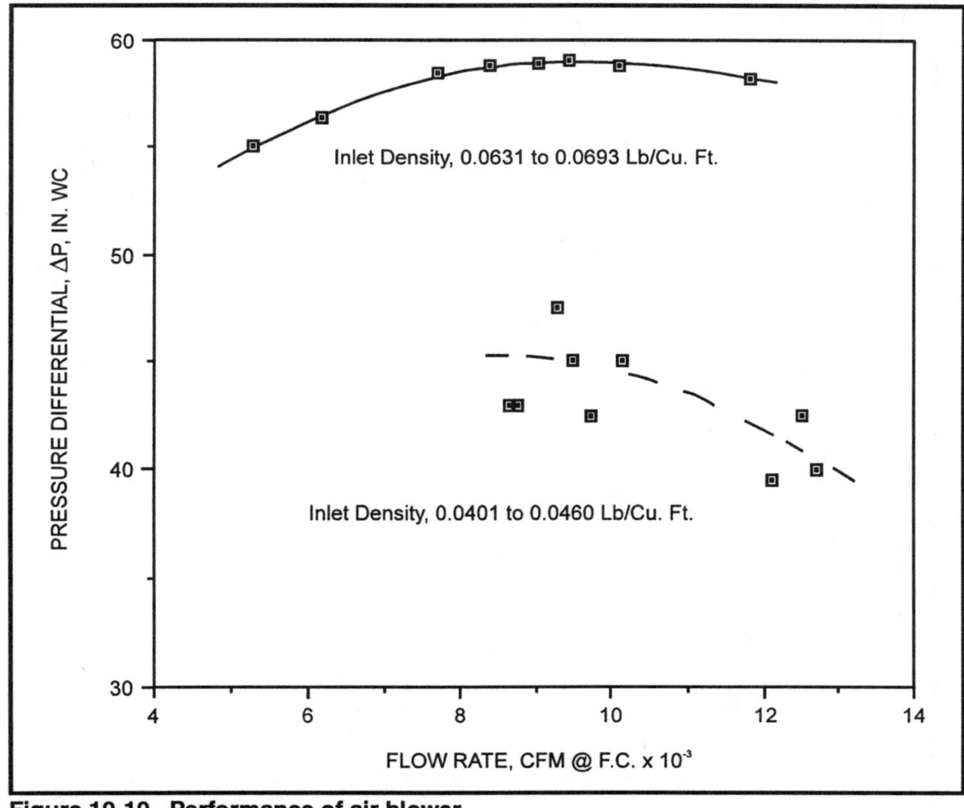

Figure 10-10. Performance of air blower

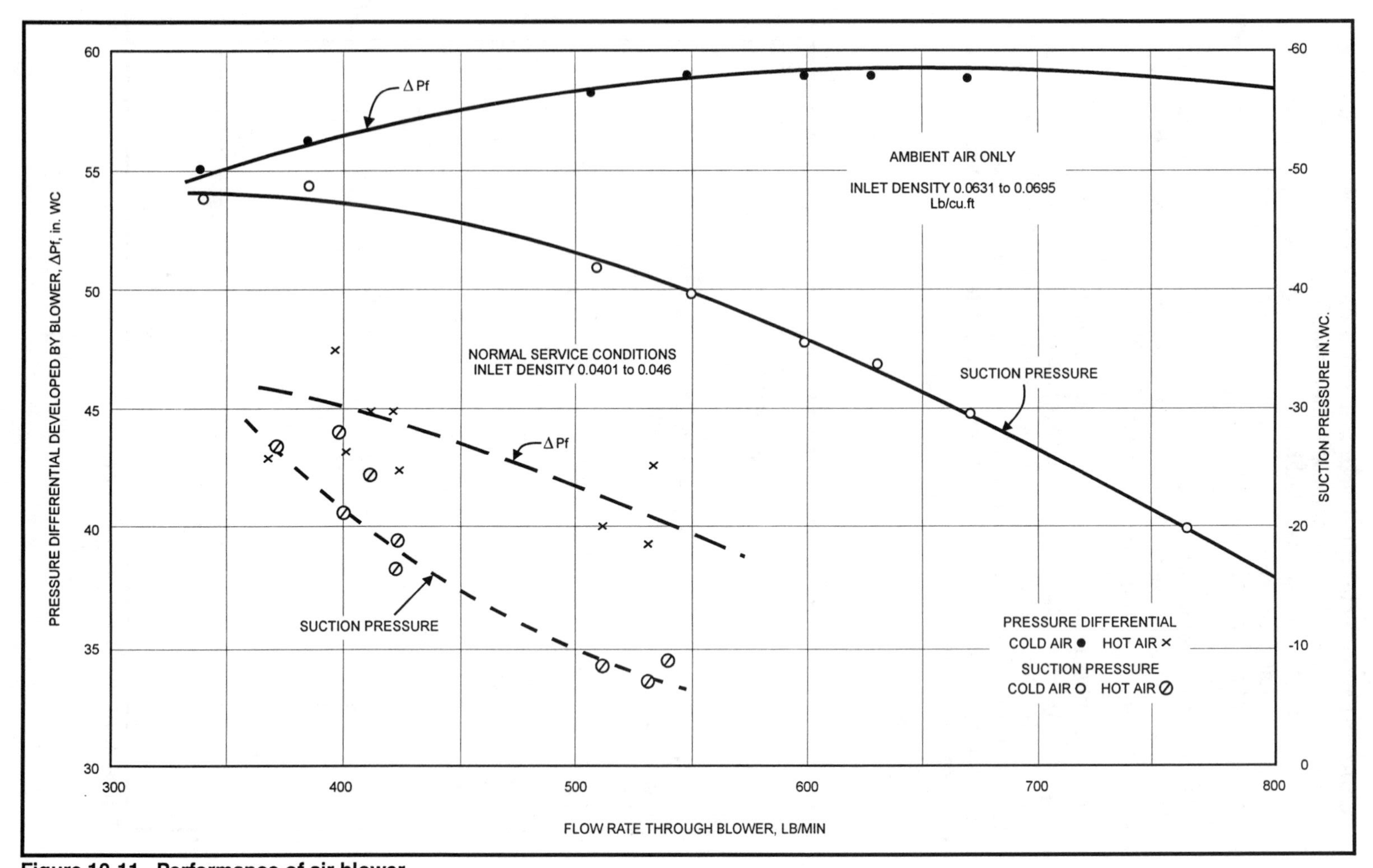

Figure 10-11. Performance of air blower

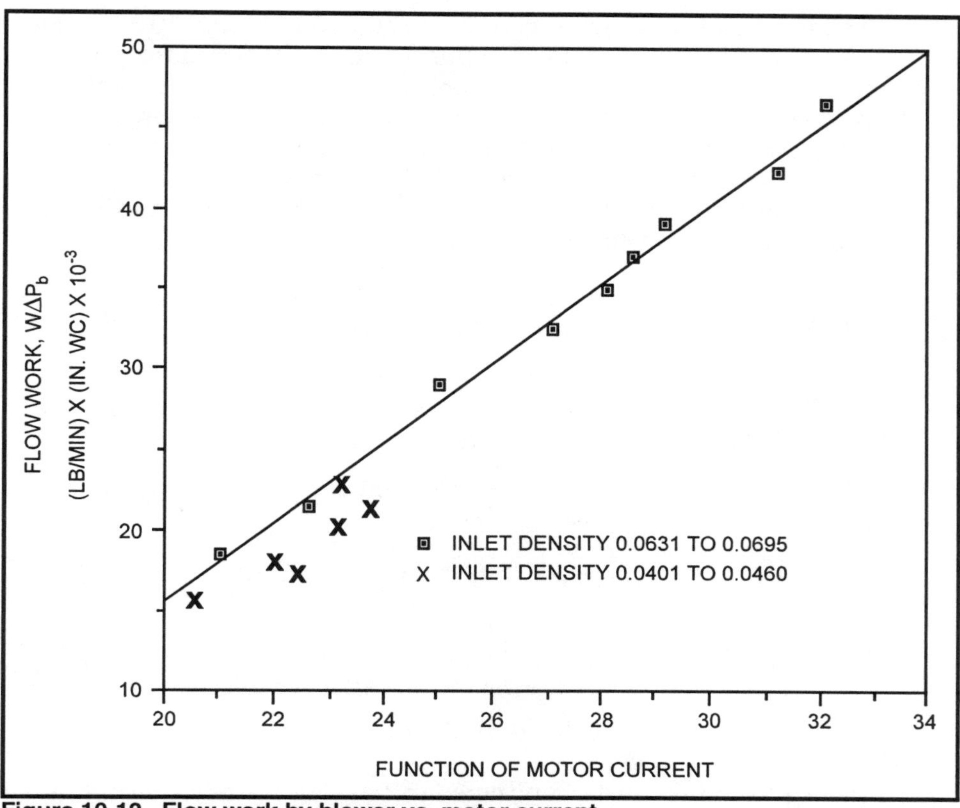

Figure 10-12. Flow work by blower vs. motor current

Procedures for Sampling Coal from the Transfer Lines

Before presenting the results of the tests of the coal mill, the technique of obtaining valid samples from the transfer lines carrying coal from the mill will be described. The probes were moved progressively along a given diameter across the 10-inch nozzle, and samples were collected at positions such that each sample would be representative of equal areas. Thus, the cross-section of the 10-inch nozzle was subdivided into four annuli and a center circle, each of the same area. The probe was then located so that samples would be taken at the midpoint of each area on either side of the center.[3] Radial positions were calculated from the relation

$$\%_r = \frac{2n-1}{N}, \text{ where } n = 1, 2, 3 \ldots\ldots \frac{N}{2}$$

where N = number of sample points, in this case 10.

3. *Chemical Engineers' Handbook,* 3rd Ed., p. 399. McGraw-Hill Book Co., Inc. (1950)

The results of two such traverses are shown in Figure 10-13, where it is seen that the E-W traverse exhibits a remarkable symmetry, while the N-S traverse is slightly skewed. An analysis of the data showed that samples collected at positions 4 and 7, for both traverses, represented 17.5% of the total. Since the probe was positioned at points along each radius so that five equal areas were sampled, the total flow rate of coal could be computed from the measurements made at positions 4 and 7, from the following relationship:

$$W_c = \frac{1}{2}\left[\left(\frac{W}{\Delta T}\right)_4 + \left(\frac{W}{\Delta T}\right)_7\right] \times \left(\frac{20}{17.5}\right)\left(\frac{10}{0.5}\right)^2\left(\frac{1}{454}\right) = 0.505\left[\left(\frac{W}{\Delta T}\right)_4 + \left(\frac{W}{\Delta T}\right)_7\right]$$

where

W_c = lb. coal flowing per minute

$\left[\left(\frac{W}{\Delta T}\right)\right]_{4,\,7}$ = $\frac{\text{weight of coal sample, gm}}{\text{sample time, minutes}}$ at probe positions 4 and 7

$\left[\left(\frac{10}{0.5}\right)\right]$ = ratio, $\frac{\text{transfer line nozzle diameter}}{\text{probe diameter}}$

$\left[\left(\frac{20}{17.5}\right)\right]$ = ratio, $\frac{\text{theoretical \% of total in a given area}}{\text{actual \% of total withdrawn in area}}$ represented by positions 4 and 7

In all subsequent tests, conducted under steady-state conditions, samples were withdrawn from positions 4 and 7 from each transfer line, and the rate of coal transported calculated as indicated above. This assumes the same percentage of coal appears at these points at all conditions of mill operation, a compromise made in the interest of time.

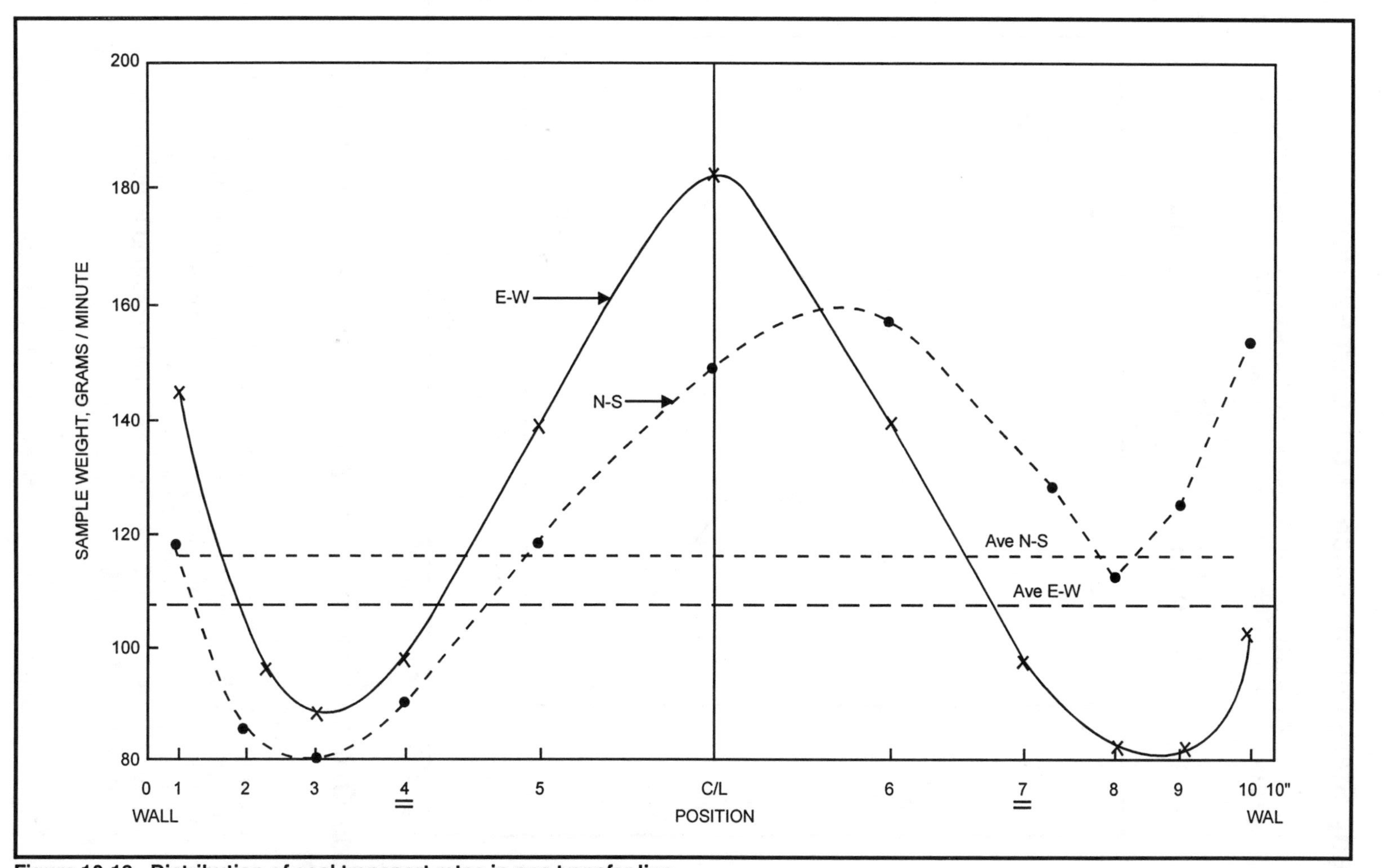

Figure 10-13. Distribution of coal transport rates in one transfer line

Steady-State Performance of the Coal Feeder

Samples were withdrawn in the manner described, while the mill was operated at seven different values of air and coal flow rates. The range of coal feed rates is indicated in Figure 10-14, which also shows the relation between measured coal rates and feed screw speed. Little confidence was placed in the screw feeder as a constant delivery mechanism, however.

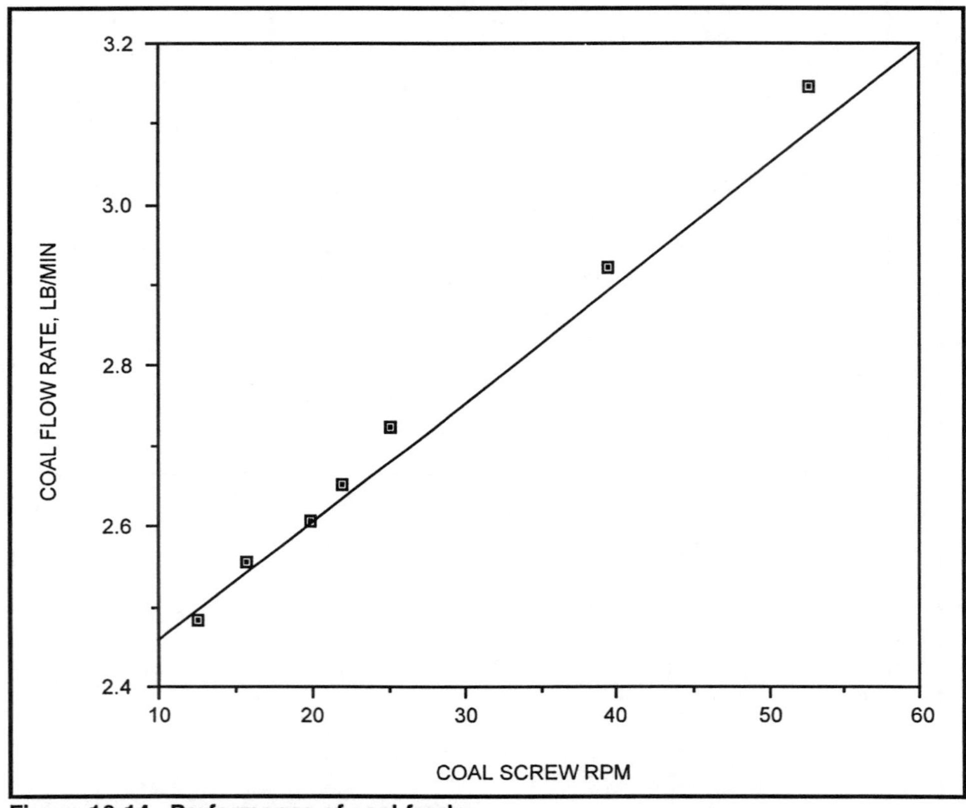

Figure 10-14. Performance of coal feeder

Nozzle-Orifice Assembly for Measuring the Flow Rate of Coal

Methods of measuring simultaneously the flow rate of gas and coal transported through pipes have been described in the literature.[4] One such device consists of a flow nozzle and an orifice plate in series, separated an appropriate distance from one another, through which the mixture flows. The mixing orifice and nozzle installed in the system under study fulfilled these requirements, although the positions were reversed.

4. "Meter for Flowing Mixtures of Air and Pulverized Coal," H. M. Carlson, P. M. Frazier, and R. B. Engdahl. *Trans. ASME*, Feb. 1948, pp. 65-79.

Nevertheless, since pressure drop and pressure measurements were available, the concept was tested. The results, shown in Figure 10-15, were most gratifying.

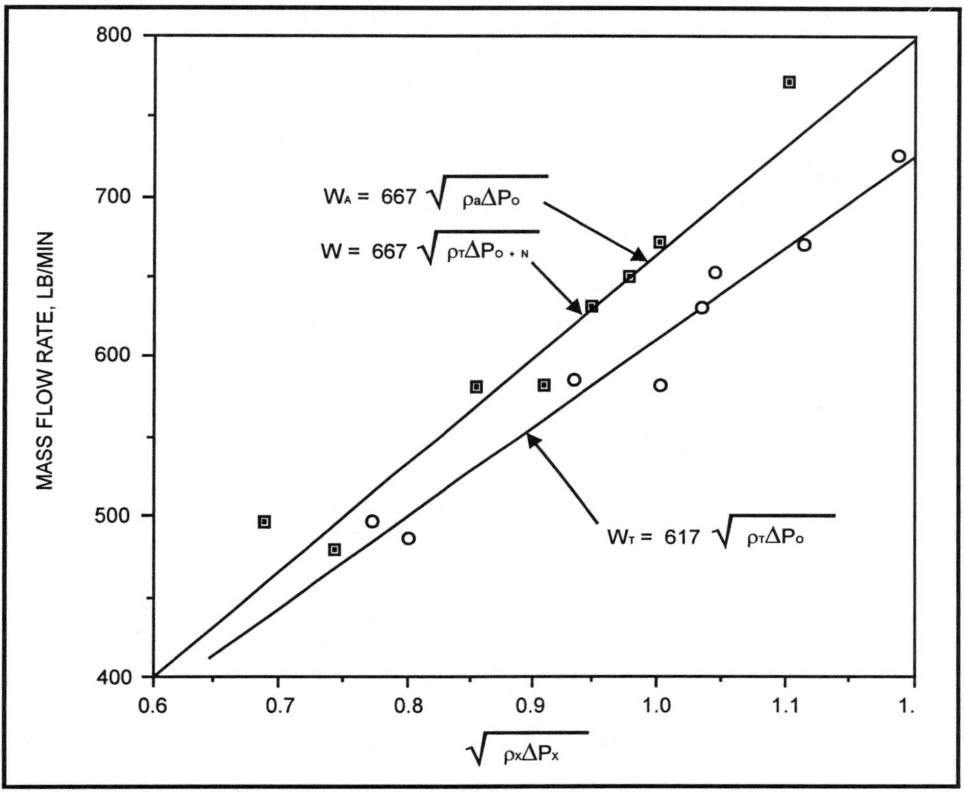

Figure 10-15. Coal flow rates from pressure and pressure gradients

The lower line in Figure 10-15 was derived from the measured pressure drop across the 9-inch orifices in the mixing sections; the upper line, using the pressure differential between orifice fore pressure and nozzle throat pressure. (By coincidence, the upper line corresponds to that calculated for the orifices when air only flows and a discharge coefficient of 0.61 is assumed). The equations of the two graphs are:

$$W_T = 667\sqrt{\rho_T \Delta P_{o+n}}, \text{ air plus coal}$$

$$W_A = 667\sqrt{\rho_a \Delta P_0}, \text{ air only}$$

$$W_T = 617\sqrt{\rho_T \Delta P_0}, \text{ for the lower line, air plus coal}$$

The inference can be made that, in order to use an orifice plate as a meter for these air and coal mixtures, a flow coefficient of about 93% of that used for air only is required. How these correlations may be used to estimate the flow rate of coal in the transfer lines is illustrated below.

Using the 9-inch orifice data, and assuming the flow rate of air, W_A, is known,

$$W_T = 617\sqrt{\rho_T \Delta P_0}$$

$$W_T = W_C + W_A$$

where W_A and W_C are the mass rates of flow of air and coal, respectively. We assume an independent measure of the flow rate of air is available.

The density of the mixture is $\rho_T = \dfrac{W_C + W_A}{Q_A} = \dfrac{W_C + W_A}{W_A/\rho_A}$

Therefore, $(W_C + W_A) = 617\sqrt{\dfrac{(W_C + W_A)}{Q_A}\rho_A \Delta P_0}$

from which $(W_C + W_A) = (617)^2 \rho_A \left(\dfrac{\Delta P_0}{W_A}\right)$

If the molecular weight of air is 28.85, then $\rho_A = 0.054\dfrac{P_0}{T_0}$, so that

$$(W_C + W_A) = (617)^2 (0.054)\frac{P_0}{T_0}\left(\frac{\Delta P_0}{W_A}\right) = 20,550 \cdot \frac{P_0}{T_0} \cdot \frac{\Delta P_0}{W_A}$$

Since the flow rate of air is assumed known, the flow rate of coal follows directly.

Figure 10-15 shows that if the orifice plus the nozzle pressure drop, $(\Delta P)_{o+n}$, is used, the relationship becomes

$$W_C + W_A = 667\sqrt{\rho_T (\Delta P)_{o+n}} = 24,050\frac{P_0}{T_0}\frac{(\Delta P)_{o+n}}{W_A}$$

Figure 10-16 may be used to solve this equation by using a fictitious value of P_0/T_0 17% greater than that measured. That is, if the value of P_0/T_0 is multiplied by 1.17, Figure 10-16 may be used to find the coal flow rate using the pressure drop measurement across both the 9-inch orifice and the 10-inch nozzle.

Thus $(W_C + W_A) = 20{,}550\ (P_0/T_0)_f \cdot (\Delta P)_{o+n}/W_A$

where $(P_0/T_0)_f = 1.17(P_0/T_0)_{abs}$.

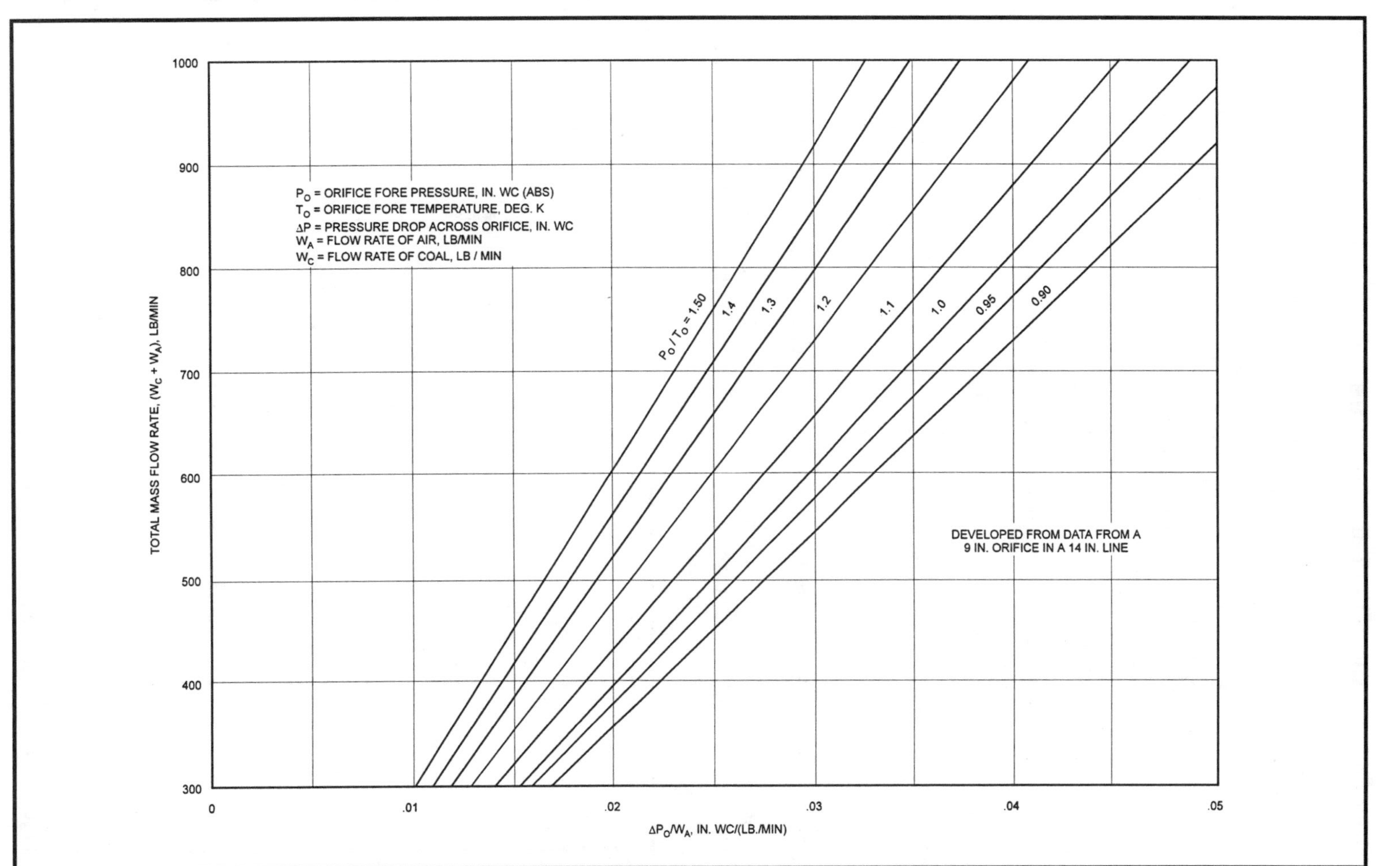

Figure 10-16. Graph for estimating the rate of pulverized coal flowing through a 9" orifice in a 14" pipe

Coal Flow Rates from Mill Measurements

The above results suggested that perhaps measures of pressures and pressure gradients across the mill might be correlated with the flow rates of air and air-coal mixtures. (The result is shown in Figure 10-17.) The density of air at the mill inlet was used in all cases. Pressure drop measurements, across the mill alone, were found to correlate with the sum of air and coal rate flowing through it, in which case the following relations may be used directly:

$$W_C + W_A = 614\sqrt{\rho_{mi}(\Delta P)_{mi}}$$

and using the same molecular weight as previously,

$$W_C + W_A = 614\,(0.054)^{1/2}\,[\,(P/T)_{mi}(\Delta P)_{mi}]^{1/2}$$

or

$$W_C + W_A = 143\,[\,(P/T)_{mi}(\Delta P)_{mi}]^{1/2}$$

Figure 10-18 was derived from the experimental data.

The practical range of values of the radical in the final relationship lies between 2 and 10. Assuming the mass flow rate of air is known, the flow rate of coal follows directly.

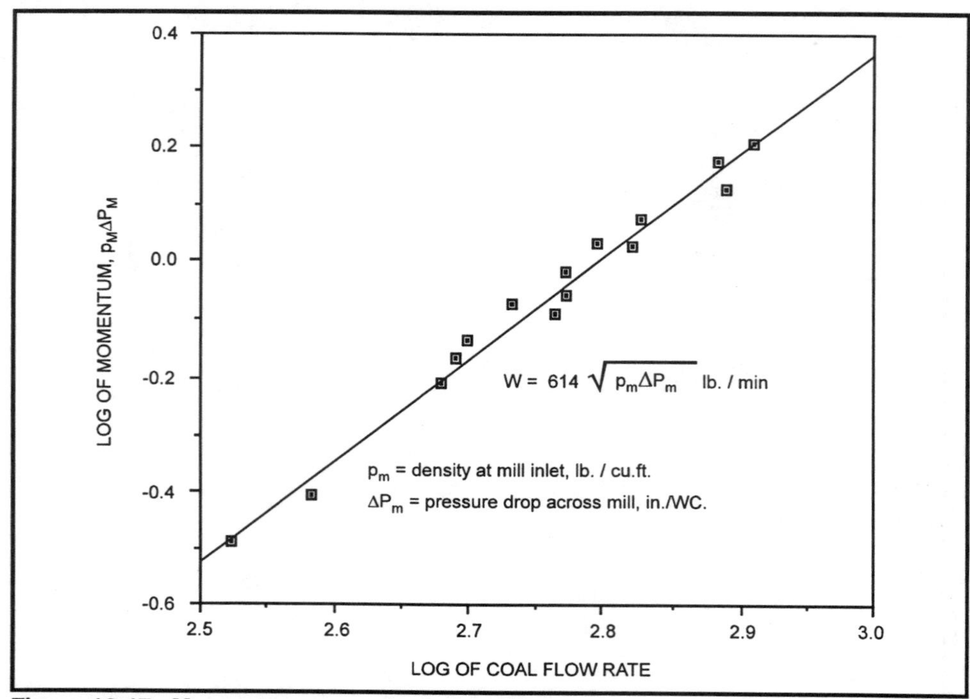

Figure 10-17. Momentum vs. coal rate for mill

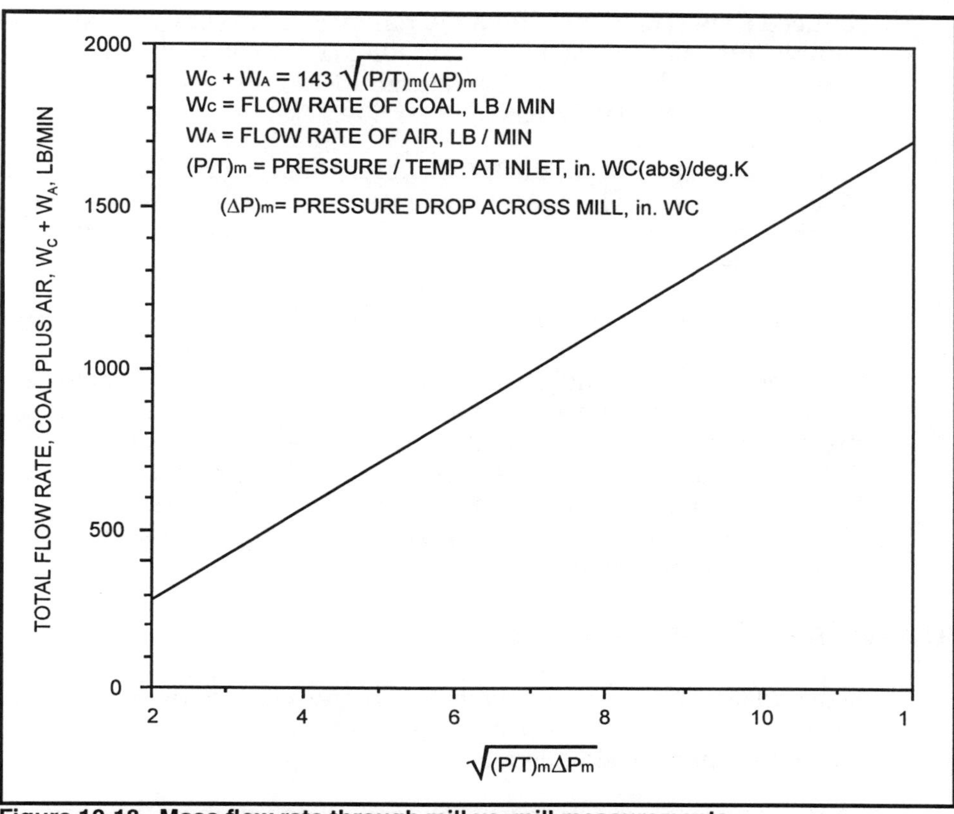

Figure 10-18. Mass flow rate through mill vs. mill measurements

Dynamic Tests of the Coal Mill

Two tests were performed to evaluate the transient response of the coal mill. The first test consisted of quickly increasing the flow rate of air to the mill. Just prior to the change and as rapidly as possible thereafter, samples were withdrawn from one of the transfer lines using the technique and apparatus previously described.

In the second test, the rate of coal addition was quickly increased, and once again successive samples were collected. In both tests the point of sampling was that position in the nozzle that had been determined as representing the average across the entire section. As expected, increasing the flow rate of air reduces the inventory in the mill and increases the rate of coal leaving. Eventually the normal output was established. The maximum increase occurred within a minute; about 2.5 minutes were required for the initial flow rate of coal to be restored.

Increasing the speed of the coal feed screw not only increases the rate of coal feed, but also the inventory of coal within the mill. However, as the inventory increases, the resistance to air flow also increases, decreasing the flow rate of air. As the inventory of coal increases, apparently even the reduced flow rate of air ultimately removes more coal, so that eventually equilibrium is again established, as long as the capacity of the mill or blower has not been exceeded. (Figure 10-19 presents the transient data obtained from the latter test.)

Figure 10-19 also shows the transient response, using the previously described correlations of coal flow rate vs. coal mill pressure and differential pressure measurements. The agreement is considered remarkable.

Analysis of the transient response data shows that the mill performs, approximately, as a first-order system with a time constant of 2.7 minutes. Obviously this feed-mill system cannot be used to compensate for rapid changes, such as can occur in the oxygen content of kiln effluent gas. If the frequency of the signal to the feeder were no greater than about 0.2 radians per minute (11 degrees per minute), the response of this mill (in the environment described herein) could be considered for this purpose, although the response of the oxygen analyzer (or other sensor) should be tested for compatibility.

Analysis of Coal Samples

Size and size distribution analyses of many samples of pulverized coal were obtained. These properties were determined by the sedimentation techniques previously described.

To determine if these properties were dependent upon the position of the sampling probe in the transfer line, samples were taken from five positions across two diameters of the nozzle. (Data are shown in Figure 10-20, where each analysis is the average of two tests.) These analyses, as well as others obtained under a variety of conditions, were considered consistent enough to assume that no appreciable difference in size and size distribution occurred at the point of sampling under any conditions existing in this investigation.

Samples were also subjected to sieve analyses and compared to data from sedimentation tests. (The results are also shown in Figure 10-20.) From all analyses it was concluded that, at all test conditions, the product from this coal mill possessed almost constant size and size distribution, including during periods of transition.

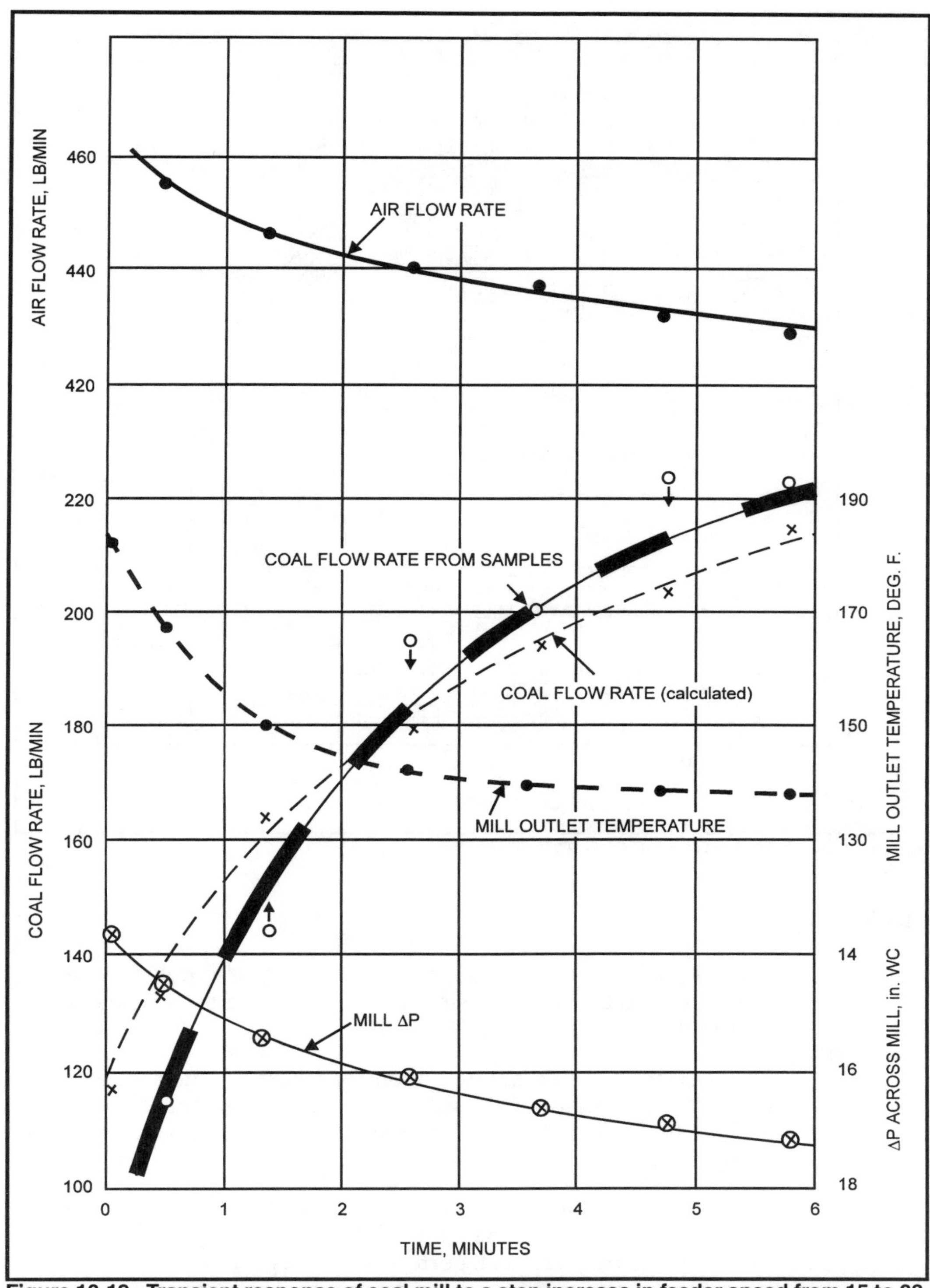

Figure 10-19. Transient response of coal mill to a step increase in feeder speed from 15 to 23 rpm

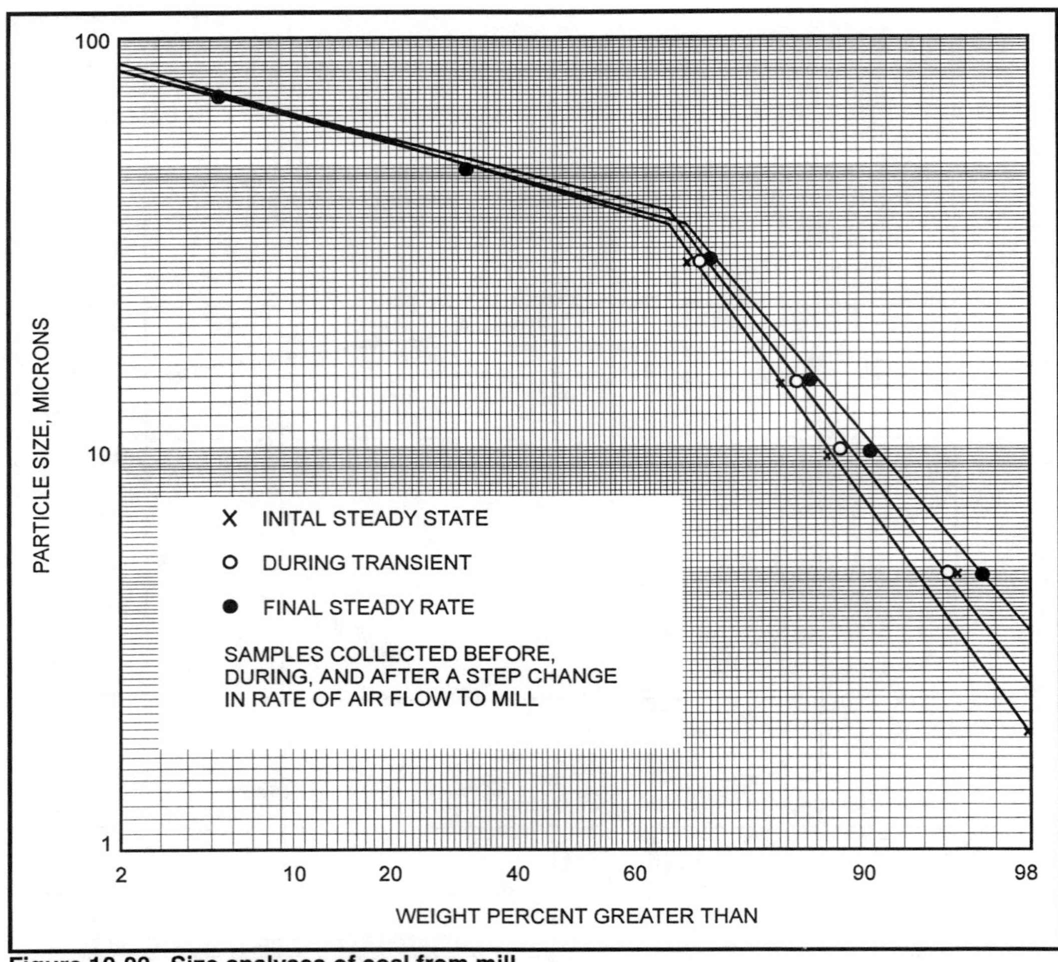

Figure 10-20. Size analyses of coal from mill

Summary

- The original processing system was obviously assembled more by intuition than by careful design, most likely because little useful information was available.
- The same may be said about the control systems, which appeared to be installed with no consideration of dynamic responses or accepted concepts of control system design. Control was almost nonexistent.
- The capacity of the blower supplying air to the coal mill was very much dependent upon the temperature of the air supplied to it. As the temperature increased, performance was sharply degraded.

- The isokinetic sampling system performed very well throughout the test program. Not only were satisfactory samples obtained when the coal mill was operating at steady state, but also during transient conditions.
- The change in the rate of output of coal from the mill, in response to an abrupt change in coal fed to it, is so slow that attempting to use the mill to regulate rapid changes in excess oxygen or other variables associated with this kiln is impossible.
- From the pressure and pressure gradient measurements associated with the orifices and nozzles in the mixing sections of the transfer lines, relations were developed from which the rate of coal transported could be found, provided an independent measure of the flow rate of air is available.
- Similar measurements associated with the coal mill can be used to obtain the rate at which coal is processed, again provided the flow rate of air through the mill is known.
- The size and size distribution of the coal from the mill remains virtually unchanged despite changes in coal feed rate and flow rate of air.

Facts Essential for Devising a Control Strategy

The purpose of this process was to transform a granular solid having the consistency of fine sand into a lightly calcined product that could be crushed to granules about the size of peas. In this form the material could be readily handled, mixed with other solids, and efficiently processed without undue loss.

The desired transformations occur within the kiln where temperatures and residence time should be sufficient to accomplish the mild calcination and yet avoid creating large fused aggregates, which are difficult to handle and to reduce to appropriate dimensions.

Rather stringent conditions were imposed on operating procedures, conditions that may be impossible to attain consistent with reality. These are stated below; the reader will promptly recognize the conflicts and problems posed.

First, the rate and, perhaps also, the size distribution and quality of the ore-bearing solid material fed to the kiln were subject to change and might vary in physical or chemical composition. More importantly, it was virtually mandatory that the byproduct gas be used as the primary fuel, augmented by coal and natural gas. Since the supply of the by-product fuel was variable, the composition of the composite fuel was also variable. Thus, steady-state conditions within the kiln would be impossible to sustain over long periods of time.

In the interests of energy conservation, the use of hot air from the product cooler was desirable. However, the performance of the blowers deteriorated at higher temperatures, so heat conservation by this means is limited. The operation of the coal mill system was also dependent upon both the temperature and flow rate of the air supplied to it.

The velocity of air flowing through the transfer lines leading from the mill to the kiln must be sufficient to produce a uniform suspension of coal entering the kiln.

Leakage of air into the kiln would be difficult to control, rendering analyses of stack gas of questionable utility.

Carryover of dust both from the stack and in the hot recycled air may also create problems.

Recommendations

The following changes in processing and operating procedures for such systems as described above are recommended.

As currently done, take hot air from below the crushing rolls at the kiln, but following the dust separator, and add this, at measured flow rate to: (1) the gaseous fuel prior to the kiln, and (2) the transfer lines carrying coal from the mill, bearing in mind that a velocity of about 80 feet per second is required to ensure a uniform suspension in the coal transfer lines.

Control the hot air admitted through valve B so that the temperature of air entering the coal mill blower will remain in an acceptable range.

Measure the flow rate and temperature of air to the mill so that, together with pressure and pressure gradient measurements from the orifices and nozzles installed in the transfer lines, the flow rate of coal can be computed and displayed. The correlations similar to those developed herein can be used for this purpose.

Assuming a measure of the inventory of by-product gas is available, maintain this inventory constant and change the flow rate of coal to the kiln to compensate, provided the rate of changes in the flow rate of by-product gas does not exceed the capability of the coal mill to respond.

In addition, use coal feed rates to adjust temperatures within the kiln, since flame temperatures and temperature distribution may respond more favorably to changes in the rate of this fuel. This assumes favorable dynamic response from the coal mill.

Also vary the excess air to assist in establishing favorable temperatures within the kiln. This must be accomplished by trial and error, observing the quality of the product. Any measures of temperature within the kiln would be highly suspect.

When the use of natural gas is necessary, make provision for increasing the flow rate of air to compensate for the greater demand. Adjustments of kiln inlet pressure may be necessary to ensure increased air intake at the burners.

Manipulate the "valve" at the kiln exit to ensure a satisfactory pressure at the firing end of the kiln. Together with the previous two recommendations, the terminal valve can be used to assist in establishing desirable kiln conditions.

Burners should be designed to safely handle the wide range in fuel-to-air ratios and flow rates, an item outside the scope of this study.

If stack dust creates an environmental problem, the installation of an electrostatic precipitator could be required.

Final Comments

There appears to be no practical way to minimize excess air, in fact it may be counterproductive to do so, since an optimum temperature profile within the kiln may require air in excess of that theoretically required for the combustion of the fuels. After all, the main objective is to prepare a processed solid with desirable characteristics, the price for which may be sacrifices in heat conservation.

While there probably exists an optimum mix of fuels that can produce an acceptable product at lowest cost, this may never be discovered until a stable and controllable operation is established. At a given kiln roll rate and feed of constant quality, for each combination of fuels there may be an ideal rate of firing. These unique conditions can only be sought by careful experimentation. In view of the complicated interactions among many variables, the major ones being largely unpredictable, these conditions may never be discovered.

Although this system may be considered a candidate for multivariable control, very likely this degree of sophistication is not needed. The "time constants" of the components in each loop of the proposed arrangement are widely different. Thus, of the variables readily measured with highly responsive sensors, pressure in the kiln is the most rapid, gas and air flow rate changes can be made moderately rapid; while alteration in the quality of kiln product is certain to be the least responsive. The separation of "time constants" among the different subsystems is probably sufficient to enable each control loop to be designed as a single-input, single-output system, provided adequate steady-state and dynamic information is on hand.

Product quality is the one output the measure of which is most difficult to obtain. Considerable ingenuity will always be required to devise satisfactory means of procuring such a measure, an ever-present problem encountered when attempting to monitor the quality of product from fired kilns.

From this case study it is apparent that extreme effort is sometimes required to create some kind of order out of a chaotic situation. Unfortunately, we are denied the satisfaction of reporting the extent to which the recommendations were considered, what action was taken, or the results achieved. A casual reader cannot, however, fail to note that a well-executed experimental study can yield useful and sometimes unexpected results, which can be used as a basis for the logical design of processes and control systems for seemingly intractable operations.

APPENDIX A
PCPARSEL: A Program for Retrieving Optimum Controller Parameters

At several points in this book recommended parameters are given for controllers associated with systems for which dynamic characteristics had been determined. These parameters were obtained from correlations embedded in the software program entitled PCPARSEL. At one time this software was distributed by ISA. It is no longer available from that source.

PCPARSEL was developed by the author with the hope of encouraging experimental determination of system dynamics, since once having this information, optimum controller parameters would be immediately available from the program. This procedure, based on known open-loop system response, seemingly would be preferred to common practices of relying on "tuning" techniques or "self-tuning" routines. In response to requests, a brief resume of this program is presented here.

An interacting computer program was prepared by which optimum controller parameters could be found for a given system, with the synthesis carried out in the frequency domain by trial and error.

On the screen was presented the Bode diagram of the assumed process, followed by the amplitude and phase diagrams for the selected controller algorithm with all gains unity. At this point some adjustments of controller parameters could be made if desired.

Next the closed-loop diagrams were presented, and if the closed loop frequency response was not satisfactory, the user could return to an earlier presentation and adjust the controller parameters. The closed-loop gain was determined by shifting the combined function, with unity gain, on a Nichols diagram (appearing on the screen) until tangency to the 2 db closed-loop locus was achieved.

The selection of optimum controller parameters was achieved when the following criteria were satisfied:

1. For those closed-loop systems that can exhibit a maximum amplitude greater than unity, the frequency at which the amplitude becomes unity should be maximum. This frequency shall be called the crossover frequency.
2. The dynamic amplitude ratio of the closed loop shall have a maximum of 2 decibels. Again, this applies to those closed-loop systems that can exhibit a maximum.
3. At frequencies less than ω_{co} (crossover frequency), the dynamic amplitude ratio shall always be unity or greater for those closed-loop systems that exhibit a maximum.
4. In addition, to avoid undesirable conditions, which are unique to some systems, in no case must the dynamic response angle of the fully compensated open-loop function become less than –128 degrees before it assumes a monotonic decline toward and beyond –180 degrees. Nor shall the open-loop phase angle ever become more negative than 150 degrees at a frequency less than that at which the open-loop amplitude ratio becomes less than unity. This latter rule allows for a 30-degree phase margin.

The "processes" selected included pure integrator, first order, second order, and third order, with a wide selection of time constants and delay time varying from zero to a value equal to the major time constant of the process.

The choice of controller algorithms were:

1. Proportional, K
2. Proportional plus integral, $K(1 + T_i s)/T_i s$
3. Proportional plus integral plus "derivative,"

$$K\frac{(1+T_i s)}{T_i s}\frac{(1+AT_d s)^n}{(1+T_d s)}$$

with $A = 10$ and 30 and $n = 1$ and 2.

Correlations were then found relating system parameters and optimum controller parameters for the selected controller algorithm.

A unique feature of this software is that the crossover frequency for each optimum combination is presented. This is the frequency at which the amplitude of closed-loop function crosses unity, for those closed-loop systems that possess a maximum amplitude. The crossover frequency is a convenient measure of the closed-loop performance of the final combination of system plus controller.

The program has considerable pedagogical value since it can be used to study the deleterious effect of delay time (dead time), the enhancement produced by using "derivative" action, and the problems encountered when using integral control with a process, which is itself an integrator.

APPENDIX B Frequency Response from Pulse Test Data Using the Fourier Transform

Recovery of frequency response information from pulse tests has been extensively described (see references 1 to 11) and, hence, need not be considered in detail here. Better programs and improved computation and data processing facilities are available today than at the time of these studies. Nonetheless, certain features of yesteryear's computer program may merit presentation, features that are particularly convenient, and quite necessary, when processing data derived from plant tests.

In Ref. 1, the trapezoidal approximation of the Fourier transform computer routine is described, and the frequency range over which reliable dynamic characteristics of known systems can be obtained is amply demonstrated. This study also shows that, for input functions possessing zero amplitudes at certain frequencies, reliable results can be derived at frequencies adjacent to and between these specific frequencies. While actual practical input pulses very rarely contain zeros in their spectra, computed real and imaginary parts of inputs, outputs, and the ratios can approach zero. Thus, it is important that any practical Fourier transform should print out both the real and imaginary parts of inputs and outputs at all frequencies for which results are requested. This information is valuable in indicating where uncertainties in both amplitude and phase angle can occur. Other practical requirements to be considered are discussed below.

While input pulse forcing functions have identical initial and final values, outputs may not always return to their initial state. Two reasons exist for this behavior: (1) small, possibly inadvertent or unknown, changes occur between input and output, which cause departures from the initial state, and (2) the process itself may act as an integrator.

For these reasons the computer program must be capable of recovering the frequency response information from non-closing outputs.

To accommodate the first situation, there are several alternatives. The most obvious is to adjust all output data to a new bias extending from the initial to final value of the records. Another technique is to merely smooth the records beginning at some point near the end of the output. Or the output records may be adjusted by assuming an exponential decay starting at some point near their terminal points. These techniques appear to work quite well if the output has a slowly decaying termination. Another alternative is to merely truncate the output at some logical point. (See reference 3 below). Considerable experience is required to gain confidence in these simple data adjustment methods, and it is suggested that novices experiment with known systems to determine the overall effect on results obtained.

To accommodate the truly non-closing output pulse, the computer routine is slightly altered. This results in using the derivative of the time histories, which always close. We show here the two transformations:

The Fourier transform for extracting frequency response information from pulse test data

$$\hat{PF}(s) = \frac{\int_0^{\infty} g(t)\, e^{-st}\, dt}{\int_0^{\infty} f(t)\, e^{-st}\, dt} = \frac{\int_0^{\infty} g'(t)\, e^{-st}\, dt}{\int_0^{\infty} f'(t)\, e^{-st}\, dt}$$

When a pulse output does not close (return to its original value) the following modification of the transform is used.

$$\hat{PF}(s) = \frac{\int_0^{\upsilon T_2} e^{-st} g'(t)\, dt}{\int_0^{\mu T_1} e^{-st} f'(t)\, dt} = \frac{e^{-s\upsilon T_2} g(\upsilon T_2) + s\int_0^{\upsilon T_2} e^{-st} g(t)\, dt}{e^{-s\mu T_1} g(\mu T_1) + s\int_0^{\mu T_2} e^{-st} f(t)\, dt}$$

A practical Fourier transform computational routine should possess all these features.

The program used by the author required dividing pulses along the time axis into increments of a specified size. Two sizes could be used, one to resolve rapidly changing time histories and larger increments in the usual slowly decaying portion of output response records. With modern data handling facilities this is probably unnecessary.

In addition, an option was available so that results could be normalized, that is, at zero frequency the amplitude of the ratio was made unity. In any event, one should always choose zero as one frequency at which the transform is computed. The ratio of the area under the output divided by the area under the input gives an estimate of the steady-state ratio of output to input, or process gain.

The user is left to his own devices to derive the linear approximation of the "transfer" or frequency function from the frequency response computation. Actually, this is a rather simple procedure and can be accomplished by plotting the frequency response data as a Bode diagram and reconstructing the experimental curves using combinations of first- and second-order forms, the latter with various damping ratios and all having the same scale. By judicious choice of the theoretical forms a combination may be found that will agree fairly well with the experimentally derived Bode plot.

References

1. Draper, C. S., Walter McKay and Sidney Lees. Instrument Engineering, Vol. 2, McGraw-Hill Book Co., 1953. Chapter 25, "Relating Function Forms from Pulse Function Responses."

2. Dreifke, Gerald E. "Effects on Input Pulse Shape and Width on Accuracy of Dynamic System Analysis from Experimental Pulse Data," Dissertation presented to Sever Institute of Washington University, St. Louis, Mo., June, 1961.

3. Dreifke, G. E., G. Mesmer and J. O. Hougen. Effects of Truncation on Time to Frequency Domain Conversion. Trans. ISA 1, No. 4 (October 1962), pp. 353-368.

4. Hougen, J. O., Experiences and Experiments with Process Dynamics, AICE, Monograph Series, No. 4, Vol. 60, 1964.

5. Dreifke, G. E. and J. O. Hougen, "Experimental Determination of System Dynamics by Pulse Methods." Fourth Joint Automatic Control Conference, Minneapolis, MN (1963).

6. Dreifke, G. E., "Effects of Input Pulse Shape and Width on Accuracy of Dynamic System Analysis from Experimental Pulse Data." Dissertation presented to the Sever Institute of Washington University in partial fulfillment of the requirements for the degree of Doctor of Science, St. Louis, Mo., 1961.

7. Dreifke, G. E. and J. O. Hougen, "Process Diagnosis and Model Formulation by Pulse Methods." Instrumentation in the Chemical and Petroleum Industries, Volume 2. Proceedings of the chemical and petroleum instrumentation sessions held during the First Joint International Symposium on Analysis Instrumentation and Chemical and Petroleum Instrumentation, May 26-28, 1965, Montreal, Canada. A publication of ISA. pp. 81-113.

8. Hougen, J. O. and R. A. Walsh, "Pulse Testing Methods," Chemical Engineering Progress, Vol. 57, pp. 69-79 (1961).

9. Hougen, J. O., "Experiences and Experiments with Process Dynamics," Chemical Engineering Progress, Monograph Series, No. 4, Vol. 60, (1964). AICE, 345 E. 47 St., New York, NY.

10. Banham, J. W. Jr., "Experimental Determination of Open-Loop Frequency Response Characteristics of DLG-9 Class Steam Generator Systems," A report of Naval Boiler and Turbine Laboratory, Philadelphia, PA. (1964).

11. Banham, J. W. Jr., "Development of Experimental Techniques for Frequency Response Analysis by Pulse Test Method," Report of Naval Boiler and Turbine Laboratory, Philadelphia, PA. (1964).

Readers who may be interested in methods of determining process dynamics by experimental methods are advised to consult more recent literature.

APPENDIX C
Data Systems for Plant Testing

Suppose one wishes to measure the change in pressure of superheated steam at 600 psi. The usual complement of industrial instrumentation might employ a sensor-transmitter with a range from 200 to 700 psi (this range selected to encompass pressures anticipated during process start-up to the expected maximum). Figure C-1 shows the output from an industrial transducer/transmitter going from 4 to 20 ma over the above excursion of pressure.

Next suppose that the interesting information lies in the range from, say, 590 to 610 psi, such information being needed for a variety of reasons. For this purpose we wish to greatly amplify the pressure changes in the region around 600 psi. Instead of being restricted to using a signal covering a 500 psi range, we wish to look at only 20 psi. With a digital system the bit resolution is dedicated to the full range, and from this we must get the desired information covering only 4% of the range of the industrial instrument. This may not provide sufficient resolution to obtain the interesting and crucial information.

A strain-gage-type pressure transducer rated at 0 to 1000 psi with a full-range linear output of 5 mv per volt of excitation could be used for this service. Using 10 volts across the transducer bridge, the maximum output would be 50 mv, for which the sensitivity would be $50/1000 \times 1000$ or 50 microvolts per psi.

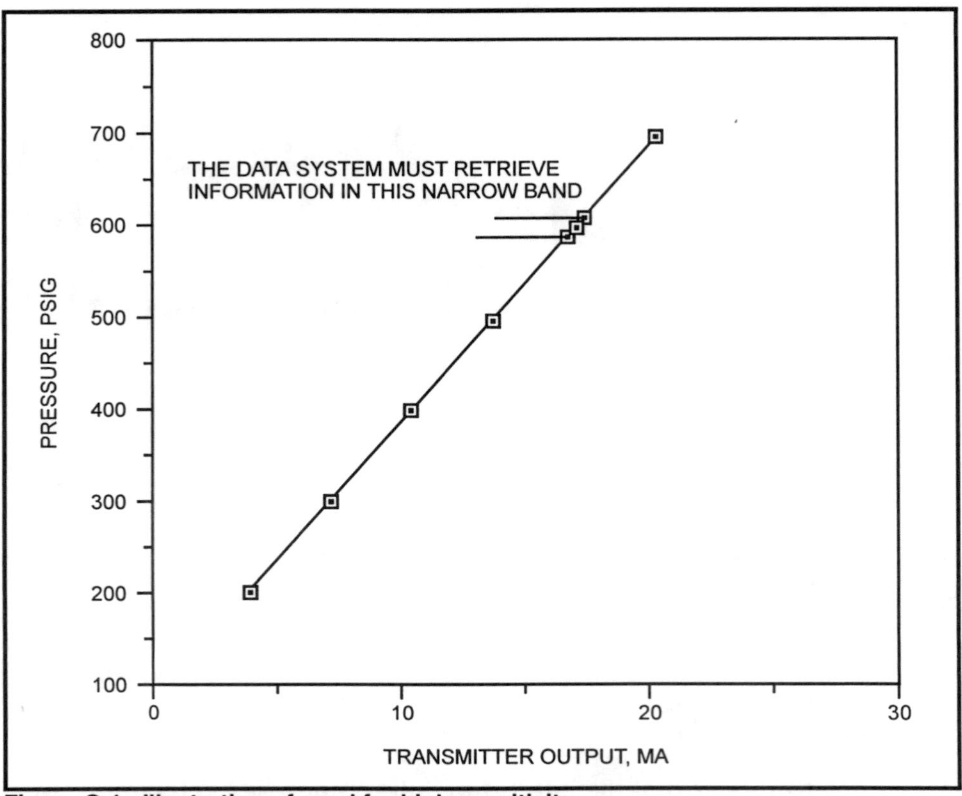

Figure C-1. Illustration of need for high sensitivity

The output signal from this transducer could be suppressed by 50 × 590/1000 = 29.5 mv, and the oscillograph recorder pen could be positioned at the lower edge of a strip chart 50 mm wide. If a pen deflection of 40 mm is desired to encompass a change of 20 psi (20 × 5 = 100 microvolts), a gain of 100/40 = 2.5 microvolts per chart division is needed. This requirement could be readily accomplished using an oscillograph recorder provided with, for example, a Hewlett-Packard low-level amplifier, Model 8803A. With proper adjustments of this recording system, changes as small as 2 psi could easily be detected. Unfortunately, H-P does not offer such an instrument today, but one would think that there must be instrumentation available from some source that could offer these features.

Instrumentation of this kind provided great capability for conditioning complex signals for the plant test work described herein. For example, if very sensitive and precise measurements of static signals were desired, the signal is suppressed so that the pen assumes a neutral position (usually in the center of the strip chart). The gain is then increased to give the desired sensitivity. (In this mode, the recorder can also serve as a highly sensitive null detector.)

Modern industrial transducer-transmitters have provisions for changing both the zero and span. Thus, there is some capability for increasing sensitivity in a selected range of the measurement. However, it may be neither safe nor feasible to do so, especially if the measurement of the variable is needed for control purposes.

For many of the studies reported here, signal recording apparatus, procured on a rental basis, was not provided with the capability described above. Although the amplifiers were adequate, suppression circuity was usually not available. Hence, the bridge circuit used for all strain-gage-type pressure and pressure differential transducers was modified as shown in Figure C-2. The small variable resistors, shown as R_{tr}, were adjusted until, with zero load on the transducer, the 10-turn potentiometer, R_7, was at its midpoint (reading of 5.000). The suppression potentiometers associated with each channel of recording, as well as the transducer, were then calibrated using dead weight testers or liquid manometers with a length of cable between transducer and data system equal to that actually installed during the testing program.

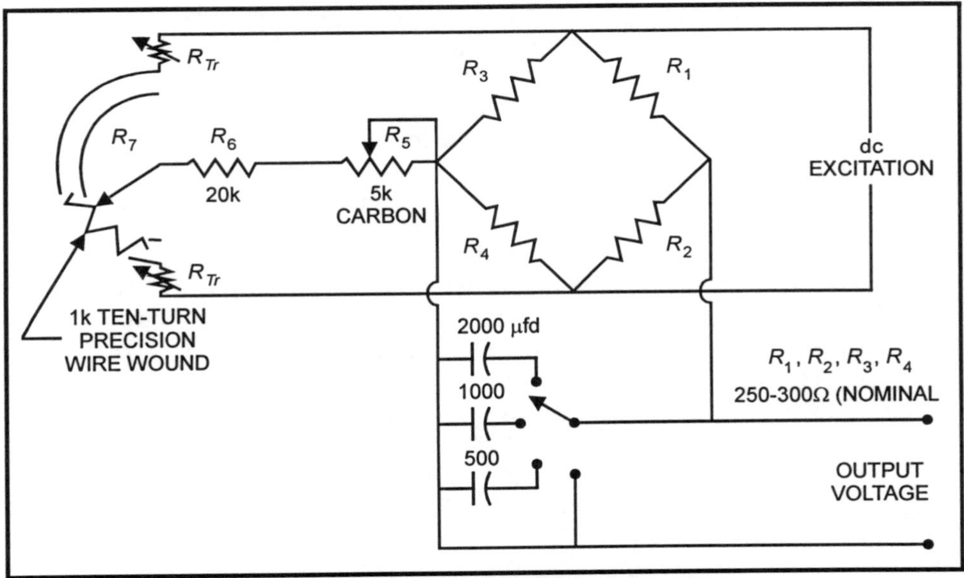

Figure C-2. Signal conditioning circuit for strain gage transducers

Temperatures of interest were measured with resistance-type sensors (RTDs) with bridge adjustment to yield 1 millivolt per degree F. Thermocouples were never used for measuring temperatures considered to be critical. The low sensitivities, about 30 microvolts per degree for iron-constantan couples, and the uncertainties between standard calibration tables and actual assemblies, were considered too great. Manufacturer's calibration data for the RTDs were usually accepted; occasionally temperature-sensing circuits were calibrated using the ice point and boiling point of water. Important temperature differences were measured directly as the difference between the output of two sensors.

A word of caution: when measuring temperatures within enclosures with sensors being able to "see" surfaces that may be at temperatures either higher or lower than the sensor itself, the errors introduced by radiation may be so great as to render the indication useless. Beware of these situations!

Signal noise, most commonly 60-cycle "hum," was removed by placing a capacitor across the dc amplifier input. As shown in Figure C-1, capacitors of 500, 1000, or 2000 μfd proved adequate in all instances.

To determine process delay (dead time) the oscillograph chart speed and sensitivity of the recorder was adjusted so that this critical information was determined with great accuracy; the time for first perceptible departure of the pen from its initial state was readily observed.

Measurement of flow rates deserves mention. For measuring large flow rates of liquids or gases, reliance on plant orifice plate installations was required, largely because of the high cost of alternative sensors. Under the very best of conditions accuracy of 5% from such installations **may** be achieved. In several instances flow nozzles were installed during a scheduled plant shut-down, but direct calibration was not possible, and reliance on manufacturer's information was required. For liquid flow rates, up to about 500 gallons per minute, turbine flow meters were used. These usually could be calibrated by measuring the rate of level change in a suitable vessel with a strain-gage-type pressure transducer. Sometimes a calibrated valve could be used for metering flow rates.

The flow loop is another device useful for making flow rate measurements. This is a section of pipe, formed into a circular section, through which the fluid to be metered flows. The loop is placed in a horizontal position with opposed pressure taps across the diameter of the conduit. The pressure differential developed is independent of viscosity, depending only on the momentum change occurring at the section across which the pressure difference is measured. This device was used on one occasion to measure the flow rate of molten salt, permitting the measurement of the heat released within an exothermic reactor.

The performance of critical control valves was always measured. These measurements included stem position, air pressure on the valve motor, air pressure to the valve positioner, and pressure drop across the valve. If available, the flow rate through the valve was also measured directly; if a meter was not installed, other means of measuring the flow rate were devised, such as measuring the rate of change in inventory in a vessel into or from which the fluid was being transferred.

Composition measurements are usually useful only for steady-state monitoring. Time for analysis, transport time between sample point and analyzer, and diffusion in the sample line must be considered. Grab sampling may be preferred in some cases.

APPENDIX D Plant Testing Protocol

Preliminary

After a problem has been identified and management has been apprised properly of the virtues of embarking on an experimental program, the following procedure is recommended. Management must be informed that some instrumentation and even process changes may be required which cannot be done except during a scheduled shut-down. Some estimate of the total cost of the experimental program should be made, realizing, however, that considerable uncertainty will exist.

The test crew, which need not consist of more than three or four people, should study the P and I diagrams, noting the process arrangement, location of all process components, and instrumentation. They should be so conversant with the process that all pertinent apparatus can be identified, *e.g.*, pipe lines, pumps, heat exchangers, vessels, and other items in the area, and know approximately the flow rates, temperatures, and pressures existing at any point in the process under consideration. Not only is the flow of material important but also the flow of energy.

Construction of block diagrams showing both the flow of material and energy are frequently useful at this stage. This exercise assists one in organizing one's ignorance.

Close inspection of all items in the process is next required. Control valves should be inspected noting their normal operating range (between 20% and 70% is a likely satisfactory range), making cursory tests for hysteresis, and observing the presence or absence of valve positioners. The installations of orifices used for measuring flow rates and sensors for temperature should be examined to determine if they comply with recommended practices. Steam traps should be functioning and not overloaded, proper pump rotation verified, and removal of all blind flanges installed during construction, and absence of hydraulic testing water in pipe loops ensured.

A list of all desired measurements is prepared. This list should be as complete as reasonable but probably will include several items that may appear unnecessary to those not involved in the testing program. Transducers are selected and calibrated and, along with auxiliary signal cable, are installed by plant personnel. The drain wires of each cable should share a common ground with the data system. All work should receive approval of safety engineers. The data receiving system should be installed in, or near, the control room where plant operators can share the observations, an arrangement which has always proved beneficial. Cordial relationships with those previously involved or presently concerned with the process to be studied is necessary to ensure appropriate consideration of the results achieved and the forthcoming recommendations.

Initial Tests

The initial observations of plant performance establishes the status quo, that is, how the process performs prior to the anticipated improvements. These also serve to identify sensitive areas and reveal unsuspected problems or malfunctions in the process or associated control systems. The data system is tested, and the personnel gains confidence in its operation and some experience with system response. These preliminary observations are usually quite important; pertinent data should be obtained and retained for future reference.

Very likely these initial tests will reveal malfunctioning or inadequate instrumentation, poorly performing valves, or improper process arrangements. Further testing must then be delayed until improvements have been made. The case study described in Chapter 1 gives examples. Within hours after the preliminary tests were started it was discovered that the major pressure disturbances were caused by variations in electric power to the furnace. Moreover, these changes were so rapid that measurement by the plant pressure sensor was impossible. In addition, the response of the control valves was found to be inadequate and the exhausters were noted to be overloaded by virtue of their recycling off-gas in attempting to execute an ill-conceived control strategy.

Testing Regimen

Steady-State Tests

Prior to conducting dynamic testing, records of the steady-state performance should be obtained. Such tests need not be time-consuming, since it is necessary only to ensure that steady state has been achieved. A few minutes of operation at a given level of operation is usually sufficient. When all variables, recorded with high sensitivity, indicate constant values, a test is completed. Some changes in operating conditions may be desired, although it may be prudent to conduct some preliminary transient tests at this point.

Dynamic Tests

With the process controllers in the manual mode, the disturbance is introduced. For the initial tests, the input should be step-like, i.e., the input should be as abrupt as practicable. Rarely can any input be truly a step, nor need it be so. This test has two purposes: (1) to determine if the process integrates, and (2) to enable rough estimates of the major time constants. The objective is to obtain some idea of the range of significant frequencies to be specified in processing pulse test data by the Fourier transform conversion program or other computational routines. These tests can also be used to determine the pure delay time between input and outputs.

It is not always feasible or necessary to place all control loops in the manual mode. For example, if unsafe conditions or existing alarms or interlocks will be activated, the control loops involved should not be opened. Usually level control loops need not be deactivated. And, of course, consultation with operators and safety personnel should always proceed testing of any kind.

Pulses used for procuring dynamic data need not be large or particularly rapid. What is desired is that the dynamics of the system be excited, but not so abruptly that the system is forced into an abnormal mode. One usually begins with small rather brief pulses, and then gradually increases the magnitude and length until acceptable outputs are obtained. Acceptable pulse length and magnitude, of course, must be determined by experience, and one tries several, some positive, others negative. The pure delay time can usually be determined from these tests, although a higher speed of recording may be required in some cases for this purpose.

Both steady-state and dynamic tests are usually conducted at several process throughputs, depending, of course, on the objectives of the test program.

No more than two weeks should be required to obtain all the data needed, usually much less, once the test crew becomes conversant with testing procedures. Such work cannot usually be carried out during the usual 8-hour increments and the test crew should expect to put in long hours, usually encompassing most of two shifts on several occasions.

Data Reduction

Procedures for computing the value of the measured variable and desired relationships should be developed so that raw observations can be converted to final results rapidly and in logical order. Usually material and energy balances are desired. Sometimes computation of reaction rates, heats of reaction, special flow rates, material and energy balances, and changes in inventory will be required. These, as well as more common calculations, can be implemented efficiently with modern computing facilities and programs.

Index